BIOREACTOR IMMOBILIZED ENZYMES AND CELLS

Fundamentals and Applications

BIOREACTOR IMMOBILIZED ENZYMES AND CELLS
Fundamentals and Applications

Edited by

MURRAY MOO-YOUNG

Industrial Biotechnology Centre,
University of Waterloo, Ontario, Canada

ELSEVIER APPLIED SCIENCE
LONDON and NEW YORK

ELSEVIER APPLIED SCIENCE PUBLISHERS LTD
Crown House, Linton Road, Barking, Essex IG11 8JU, England

Sole Distributor in the USA and Canada
ELSEVIER SCIENCE PUBLISHING CO., INC.
52 Vanderbilt Avenue, New York, NY 10017, USA

WITH 77 TABLES AND 179 ILLUSTRATIONS

© ELSEVIER APPLIED SCIENCE PUBLISHERS LTD 1988

British Library Cataloguing in Publication Data

Bioreactor immobilized enzymes and cells:
   fundamentals and applications.
   1. Immobilized enzymes
   I. Moo-Young, Murray
   547.7'58   QP601

   ISBN 1-85166-160-3

Library of Congress CIP data applied for

Printed in Great Britain by Page Bros (Norwich) Ltd

PREFACE

Bioreactors are at the heart of biotechnological processes and devices which are used in industrial, agricultural and medical applications. They rely on the bioconversion functions of agents in the form of whole living cells or cell-free enzymes which transform raw materials into useful products and/or less undesirable by-products. In recent years there has been a growing interest in bioreactors which employ 'immobilized enzymes and cells', a phrase now generally used to describe biocatalytic systems in which the bio-agents are segregated and/or attached to solid support carriers in contrast to the more conventional systems in which the bio-agents are in free suspension (cells) or are dissolved in a bulk aqueous medium (enzymes). In principle, there are many relative advantages of the immobilized systems.

This publication is a collection of papers which deals with the generic fundamentals and applications of bioreactor immobilized enzymes and cells, covering the complete range of cell types (animal, plant and microbial cells, including bacteria, yeasts and fungi), various types of bioreactor configurations and operations (fixed and fluidized beds, batch and continuous processes) and various types of immobilized techniques (adsorption, entrapment, encapsulation, compartmentalization). For convenience the material is presented according to the following three sections:

1. Techniques and principles in the preparation and characterization of immobilized cells and enzymes.
2. Applications of these systems in the production of useful products, including pharmaceuticals, foods and chemicals.
3. Applications of these systems in waste treatment and environmental pollution control.

The book is based on manuscripts which have been prepared by some of the world's foremost authorities in this area of biotechnology as well as by researchers who are becoming known as experts in certain aspects of this area. The material should be useful to students and researchers in industrial biotechnology, biochemical engineering, applied biochemistry and related fields.

In the interest of speed, this volume is produced from camera-ready copies of the original author-generated manuscripts. No attempt has been made to unify the text format in terms of such variables as literature referencing system or symbols nomenclature. In addition, only minimal editorial changes have been made to the usage of the English language in the original manuscript to improve context and clarity. (It should be noted that many of the contributors are from non-English-speaking countries.)

In closing, I wish to record my thanks to Kathy Young, one of my research assistants, who did most of the proof-reading and graphics layout of the manuscripts, and to Penny Preis, my secretary, who handled the final stages of this work. Finally, grateful acknowledgement is made to the Natural Sciences and Engineering Research Council of Canada and UNESCO for financial support of the project.

MURRAY MOO-YOUNG

CONTENTS

Section 2: Production Applications

Section 3: Waste Management Applications

LIST OF CONTRIBUTORS

S. ADACHI
Applied Microbiology, Department of Agricultural Chemistry, Faculty of Agriculture, Gifu University, Gifu 501-11, Japan

J. AMIOT
Département de Sciences et Technologie des Aliments et Centre de Recherche en Nutriton, Université Laval, Quebec, PQ, Canada G1K 7P4

L. AMOURACHE
Department of Bioengineering, Université de Technologie de Compiègne, France

G. F. ANDREWS
Department of Chemical Engineering, State University of New York at Buffalo, Buffalo, New York 14260, USA

L. A. BEHIE
Department of Chemical Engineering, University of Calgary, Calgary, Alberta, Canada T2N 1N4

H. BERNSTEIN
Department of Chemical Engineering, Massachusetts Institute of Technology, Cambridge, Massachusetts 02139, USA

A. BHADRA
Pfizer Ltd, 178 Industrial Area, Chandigarh, India

A. BORCHERT
Fachhochschule Ostfriesland, Emden, Federal Republic of Germany

E. BRAUER
Universität Frankfurt, Institut für Physikalische und Theoretische Chemie, Federal Republic of Germany

P. BRODELIUS
Institute of Biotechnology, Swiss Federal Institute of Technology, Hönggerberg, CH-8093 Zürich, Switzerland

K. BUCHHOLZ
Institut für Landwirtschaftliche Technologie und Zuckerindustrie an der TU Braunschweig, Federal Republic of Germany

R. CHAMY
Universidad Católica de Valparaíso, Casilla 4059, Valparaíso, Chile

H. N. CHANG
Korea Advanced Institute of Science and Technology, Seoul, Korea

T. M. S. CHANG
Artificial Cells and Organs Research Centre, Faculty of Medicine, McGill University, 3655 Drummond Street, Montreal, PQ, Canada

C. L. COONEY
Department of Chemical Engineering, Massachusetts Institute of Technology, Cambridge, Massachusetts 02139, USA

C. DAUNER-SCHÜTZE
Universität Frankfurt, Institut für Physikalische und Theoretische Chemie, Federal Republic of Germany

R. C. DEAN JR
Verax Corporation, Lebanon, New Hampshire 03766, USA

J. DE LA NOUE
Groupe de Recherche en Recyclage Biologique et Aquiculture, Université Laval, Ste Foy, PQ, Canada G1K 7P4

J. P. FONTA
Department of Chemical Engineering, State University of New York at Buffalo, Buffalo, New York 14260, USA

A. Freeman
Department of Biotechnology, Faculty of Life Sciences, Tel-Aviv University, Tel-Aviv 69978, Israel

G. M. Gaucher
Division of Biochemistry, University of Calgary, Calgary, Alberta, Canada T2N 1N4

K. F. Gu
Artificial Cells and Organs Research Centre, Faculty of Medicine, McGill University, 3655 Drummond Street, Montreal, PQ, Canada

T. Hayashi
Applied Microbiology, Department of Agricultural Chemistry, Faculty of Agriculture, Gifu University, Gifu 501-11, Japan

H. Horitsu
Applied Microbiology, Department of Agricultural Chemistry, Faculty of Agriculture, Gifu University, Gifu 501-11, Japan

A. Illanes
Universidad Católica de Valparaíso, Casilla 4059, Valparaíso, Chile

S. B. Karkare
Verax Corporation, Lebanon, New Hampshire 03766, USA

H. W. D. Katinger
Institute of Applied Microbiology, Vienna, Austria

H. Kautola
Biotechnology and Food Engineering, Department of Chemistry, Helsinki University of Technology, Espoo, Finland

K. Kawai
Applied Microbiology, Department of Agricultural Chemistry, Faculty of Agriculture, Gifu University, Gifu 501-11, Japan

J. Klein
Gesellschaft für Biotechnologische Forschung GmbH, 3300 Braunschweig, Federal Republic of Germany

xiv

G. KLEMENT
Institute of Applied Microbiology, Vienna, Austria

R. LANGER
Department of Applied Biological Sciences, Massachusetts Institute of Technology, Cambridge, Massachusetts 02139, USA

M. P. MARCHESE
Universidad Católica de Valparaíso, Casilla 4059, Valparaíso, Chile

J. M. MELNYK
St Bonaventure University, Olean, New York, USA

R. A. MESSING
168 Scenic Drive South, Horseheads, New York 14845, USA

H. G. MONBOUQUETTE
Department of Chemical Engineering, North Carolina State University, Raleigh, North Carolina 27695-7905, USA

M. MOO-YOUNG
Department of Chemical Engineering, University of Waterloo, Waterloo, Ontario, Canada N2L 3G1

K. NILSSON
Department of Pure and Applied Biochemistry, University of Lund, POB 124, S-221 99 Lund, Sweden

D. F. OLLIS
Department of Chemical Engineering, North Carolina State University, Raleigh, North Carolina 27695-7905, USA

Y. K. PARK
Universidade Estadual de Campinas, Faculdade de Engenharia de Alimentos (UNICAMP), Campinas 13100, SP, Brazil

G. M. PASTORE
Universidade Estadual de Campinas, Faculdade de Engenharia de Alimentos (UNICAMP), Campinas 13100 SP, Brazil

P. Prema
Regional Research Laboratory (CSIR), Trivandrum 695019, Kerala, India

D. Proulx
Groupe de Recherche en Recyclage Biologique et Aquiculture, Université Laval, Ste Foy, PQ, Canada G1K 7P4

S. V. Ramakrishna
Regional Research Laboratory (CSIR), Trivandrum 695019, Kerala, India

N. G. Ray
Verax Corporation, Lebanon, New Hampshire 03766, USA

P. W. Runstadler Jr
Verax Corporation, Lebanon, New Hampshire 03766, USA

J. M. Scharer
Department of Chemical Engineering, University of Waterloo, Waterloo, Ontario, Canada N2L 3G1

W. Scheirer
Institute of Applied Microbiology, Vienna, Austria

V. P. Sreedharan
Regional Research Laboratory (CSIR), Trivandrum 695019, Kerala, India

Y. Takahashi
Applied Microbiology, Department of Agricultural Chemistry, Faculty of Agriculture, Gifu University, Gifu 501-11, Japan

E. J. Vandamme
Laboratory of General and Industrial Microbiology, University of Ghent, Coupure Links 652, B-9000 Ghent, Belgium

K. Venkatasubramanian
Department of Chemical and Biochemical Engineering, Rutgers University, Piscataway, New Jersey 08854, USA

M. A. VIJAYALAKSHMI
Department of Food Science, Université Laval, Quebec, PQ, Canada

J. C. VUILLEMARD
Département de Sciences et Technologie des Aliments et Centre de Recherche en Nutriton, Université Laval, Quebec, PQ, Canada G1K 7P4

R. XIOA
Applied Microbiology, Department of Agricultural Chemistry, Faculty of Agriculture, Gifu University, Gifu 501-11, Japan

V. C. YANG
College of Pharmacy, University of Michigan, Ann Arbor, Michigan 48109, USA

M. E. ZÚÑIGA
Universidad Católica de Valparaíso, Casilla 4059, Valparaíso, Chile

MATRIX DESIGN FOR MICROBIAL CELL IMMOBILIZATION

J. Klein

Gesellschaft fur Biotechnologische Forschung mbH,

3300 Braunschweig, W. Germany

Introduction

Immobilized cells are now generally accepted as biocatalysts as are immobilized enzymes. This is well documented in the literature (1, 2, 3). In most cases the individual contributions can be arranged in two groups, where the first group refers to papers with main emphasis on immobilization methodology and the second group on a selected microbial biocatalytic reaction or process. On the basis of a number of papers from the authors laboratory belonging to both of the afore mentioned groups it seems justified to draw some more general conclusions on the principles of matrix design for microbial cell immobilization.

There are different routes which could be chosen to structurize this discussion. A first route could give priority to structural parameters, as summarized in Table 1.

Structural Parameters

Chemistry	Precurser Matrix	Toxicity, time scale stability regions hydrophil/phob
Crosslinking density -Porosity		Swelling Permeation, Diffusion Mechanical stability cell growth
Geometry	Size Shape	Transport processes ($-O_2$, substrates, products) -Radius/thickness -bead, fiber, block, film

Table 1: Summary of Structural Parameters Related to Cell Immobilization

1

reactor. In such a way the fluid flow of substrates and products becomes independent from the catalyst. Furthermore, in a continuous process the multiple reuse of the catalyst is automatically achieved. But also in a batch process the separation of the catalyst for reuse becomes a very simple procedure.

To achieve an optimal use of the immobilized catalytic species, i.e. the microbial cells, the geometry of the matrix is an important factor. In most cases a spherical shape will be the first choice and simple procedures have been developed to combine fast production with a control of particle diameter (6). With typical ionotropic gels, like alginates, particle diameters in the range of 3 mm down to 0.3 mm can be prepared. Controlled drying can furthermore reduce the particle size significantly and irreversibly (7).

Besides spheres, also fibrous (8) and membraneous structures can be prepared, but especially for large scale operations the use of such geometries may be limited. In general, the preparation of irregularly shaped granular particles can be avoided, i.e. by crashing of larger blocks. For a packed bed or a fluidized bed reactor a spherical shape and a controlled, narrow range of particle size is required and can be prepared today.

The retention for the immobilized cells should be as complete as possible. With the application of dead or resting cells this is achieved for most of the polymer matrices, with the exception of a small number of cells just entrapped close to the surface. The specific membraneous surface structures of ionotropic gels (alginate, chitosan) however is an integrated advantage for the prevention of cell leakage (9), since the maximum pore size of the particle surface of about 15 nm is well below the size of any microbial particle.

Cell retention becomes a problem, however, if viable and growing cells are considered. The production of ethanol with immobilized yeast or Zymomonas mobillis cells, or of biogas, are typical examples. In these processes growth of biomass is a prerequisit for the bioconversion, and the limitation of this growth in relation to product formation is one of the central problems in process control. But even at very low growth rates, due to the high productivity the biomass production is considerable. In such a case it can be advantageous to have an incomplete retention and a controlled cell leakage on such a level that the growth of new cells is balanced. To circumvent microbial growth in the fluid medium the dilution rate of the fluid phase has to be well above the washout rate of the microbes. Continuous production of ethanol with immobilized Zymomonas mobili cells can be operated in such a way for extended periods (10, 11).

In other cases, the leakage of growing cells should be completely avoided, as required for extreme product purity in food processing. A specific, cell free coating, can be used to solve this problem and simple ways have been found to prepare such coated biocatalysts. It could be demonstrated that cell growth in the medium was not possible using such catalyst preparations (8).

In a second approach the main empasis could be laid on the set of boundary conditions, which are controlling the processes of catalyst preparation and application (see Table 2)

Boundary Conditions

Cell/Enzyme	*activity*
	viability regime
Reaction medium	*composition phases*
Reactor configuration	*stirred tank*
	column
Method of immobilization	*entrapment*
	adsorption
	bonding

Table 2: Summary of Boundary Conditions Related to Cell Immobilization

In most generalizing discussions the structure is determined by the methods of immobilization (4, 5), which are usually defined as in the bottom line of Figure 1.

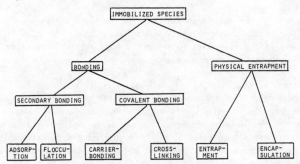

Figure 1: Mechanisms and Methods in Cell Immobilization

Looking to the literature it becomes obvious, that entrapment is by far the most widely used immobilization technique for whole microbial cells. This method is defined by a matrix formation around the cells and this has to be done in the presence of the catalytically active cells. It is of importance to realize, that such a matrix serves different functions - not just the obvious one of cell fixation - and it is this aspect of matrix function in entrapment processes which will be discussed in the following chapters.

Cell Retention

The basic principle of any heterogeneous catalysis is the immobilization of the catalytic species onto or within a solid carrier, which can easily be withheld in the

Biomass concentration

Biomass concentrations in typical bacterial suspension cultures are in the order of 100 g wet weight/L and there are limits in the biology of cell growth on one side and limits in the rheology of the suspension on the other. Cell recycle by incorporation of membranes or centrifugal separation are used to overcome the biological problems, but the rheological properties in relation to mass transport and mass transfer are a definite limit.

For the entrapment of bacterial cells biomass concentrations up to 700 g wet weight/L catalyst can be obtained and such a value can be increased by adding an additional process step of controlled drying (7). The possible upper limits depend on the type of cell and on the type of matrix. For a given matrix (epoxy carrier) the limit for E. coli cells is between 60 % and 70 % volumetric loading (12), while the corresponding value for yeast cells was in order of 30 % (13). Upper levels of biomass loading of E. coli in polyacrylamide were obtained at 15 % loading compared to 70 % in Ca-alginate (14). The rational behind these limits was a characteristic figure, related to the mechanical stability in an abrasion experiment (15). Depending on the type of polymer and the size of the organism the "structural inhibition" to the formation of a stable three dimensional matrix occurs at different levels, which however are well above the above mentioned limiting biomass concentration in suspension.

The function of the matrix in this case is therefore a way to significantly increase the volumentric biomass concentration aiming at a maximum value of catalytic species per unit volume of the bioreactor to achieve a high value of specific volumetric reaction rate and correspondingly high reactor productivities.

Maximization of biomass concentration however, is not a useful target in all cases. Especially where cells of high specific enzymatic activity are used, the limits of substrate transport are well known effects. In cases where the cost of biomass production is a significant factor, the biomass concentration has to be optimized at a submaximal level, in general, to achieve a high value of catalytic effectivity and, if possible, to work in the reaction controlled regime (5, 15) The most important parameter to be considered in this case is the substrate diffusivity, and it has been shown that the catalyst efficiency exponentially decreases with increased cell loading (16). It therefore has been shown - by calculation and by experiment - that an optimal value of cell loading is given, where a maximum value of absolute catalytic activity exist (15).

It is not appropriate at this time to repeat the strategies and findings of theoretical and experimental work directed towards the problem of catalytic efficiency for immobilized whole cells since reference can be given to the respective literature (5, 15, 16, 17, 18).

The main argument is the unique possibility which immobilization in a matrix offers, to be able to choose the appropriate biomass concentration down from very high levels with respect to the needs and the costs of a process.

Another characteristic feature to mention is the possibility of cell growth of viable cells within a given matrix. This can be used to obtain very high cell densities starting

5

from very low initial values. A factor of 10^3 for such a cell multiplication is not untypical (15). This aspect is of special importance in such cases, where morphological parameters are inhibitive to high biomass levels by direct immobilization as in the case of mycelia. Immobilization of spores and subsequent germination is an elegant way for matrix entrapment of mycelial organisms (4).

Effects of Stabilization

The stability of microbial cells in biocatalysis has to be discussed with respect to mechanical parameters on one hand and kinetic parameters on the other. While the mechanical stabilities of most microbial cells is much higher than those of mammalian or plant cells, a polymer matrix provides additional resistance against shear disruption, which might occur in a well stirred reactor. A basic requirement for such an effect is sufficient stability of the matrix itself. Based on various precurser systems polymer matrices can now be prepared, which provide considerable stability towards pressure and shear; expoxides (19) polyurethanes (20) may be mentioned as an example. For many years, in the early development phase, it was difficult to get reliable and comparable data related to mechanical stability, but due to the work towards a monograph (21) and the efforts of the working party of immobilized biocatalysis of the EFB (European Federation of Biotechnology) (22) a number of methods is available to establish a reliable set of characteristic figures for any biocatalyst preparation.

The kinetic stability of a biocatalyst is usually defined as the kinetic halflife $T1/2$, i.e. the time period in which the kinetic activity decays of 50 % to the initial activity. There is ample evidence that matrix entrapped cells show considerably higher values of $T1/2$ as compared to freely suspended cells (1, 2, 3, 7), but in most cases no mechanistic explanation can be given. The degree of stabilization and the mechanism will be different, however, if the various types of immobilized cell catalysts are considered. In this respect a) dead cells b) viable, but resting cells, and c) viable growing cells have to be distinguished. These aspects and the consequences for reactor productivities have been discussed in detail elsewhere (23). In the case of dead cells and their use in one enzyme catalized reaction, the enzyme stability within the cell is the most important factor. The matrix, however, can be of special use, if high cell density preparations are applied, which usually show a higher value of $T1/2$ than low cell density preparations. It has to be made clear that this is due to diffusional effects - at rather low levels of catalytic efficiency - and not due to an intrinsic enzyme stabilization (15).

As mentioned above, up to now no mechanistic explanation can be given to the considerable increase of $T1/2$ by matrix immobilization (24). Part of this may be due to mechanical effects, but much more likely the cell physiology is of importance. A very direct advantage of a porous matrix is the possibility for catalyst regeneration and reactivation by repeated cell growth within a matrix (24, 25). In such a way, the overall value of $T1/2$ and of the catalyst productivity can be increased considerably.

The application of viable cells with controlled growth in an immobilized cell process very much depends on the proper matrix structure, which has to give support to the viable, active cells against convective flow of the bulk fluid but which has to be porous enough for leakage of dead cells as a prerequisit to provide for replacement by newly grown cells. Alginate and carrageenan carriers are successfully applied in such processes (10, 11). This is the basis for a stationary state situation in a continuous process over time periods of months.

Microenvironmental effects

The entrapment of microbial cells in the pore space of a polymer matrix automatically leads to a segregation of the total volume into a bulk phase with convective fluid flow, and a catalyst phase, where material transport is restricted to diffusion only. In the case of diffusional limitations, gradients of concentrations and of parameters like pH are established. Consequently, the catalyst phase and the bulk phase will have a different composition and the catalytic species have an environment different from the bulk solution (17). Besides these transport controlled differences, equilibrium controlled differences can also be found, and the latter effect can be controlled by the chemical composition of the matrix. The question arises how much this specific environmental effect may be a disadvantage or an advantage and how much the advantageous effects - if proven beyond doubt - can be influenced by proper matrix design. Strong evidence for positive effects of cell immobilization has been given recently (26).

In many cases, the kinetically based concentration differences are not in favour of an improved catalytic behaviour. This is especially true where pH gradients are established leading to microenvironmental compositions not in accordance with the pH-optimum (17). Matrix design has to go towards highly porous and rather small particles to eliminate diffusional gradient as much as possible. Concentration gradients however can be favourable, if cell toxic reaction components have to be used, like oxygen or H_2O_2. While the oxidative degradation of phenole with free cells could only be achieved with oxygen concentrations given in natural air, oxygen supply for immobilized cells with oxygen enriched air became possible (24). Another example is the tolerance of the immobilized cells in sorbitol oxydation against high levels of H_2O_2 in solution (27).

Differences in the composition of catalyst and bulk phase on the basis of partition equilibria can be controlled by matrix compositon. Keeping in mind that many potential substrates for biocatalytic conversion show very poor solubility in water, efforts have been made to design porous matrices with a controlled hydrophobic/hydrophylic balance. Polyurethane carriers with specifically selected precursors have successfully been prepared, and positive effects on the transformation kinetics of steroids have been shown (29).

Other possibilities for establishing a well defined hydrophobic character are ionotropic gels, incorporating synthetic, hydrophobic ionic polymers or oligomers. The characterization of such matrices can be achieved by partition experiments with properly selected, partly or strongly hydrophobic, model compounds (29, 30). There are doubts, however, whether a hydrophobic matric is really an advantage, when water is the only solvent. The situation becomes of course different, when an organic solvent is used in the bulk phase and a two solvent system is established. In such a case the matrix should be strongly hydrophobic to support the aqueous environment for the cells.

Conclusion

In summarizing the arguments of this contribution it should have become obvious that in the context of applications of matrix entrapped microbial cells as biocatalyst there is a considerable number of effects, which are essentially controlled by the matrix structure. These aspects are compiled in Table 3.

Matrix Functions

Heterogenization/Retention		Separation catalyst/medium Residence time uncoupling
Cell Density		volumetric activity
Stabilization	kinetically, mechanically	kinetic halflife shear stability
Environment		hydrophil/phob pH, ionic strength

Table 3: Matrix Functions in Cell Immobilization

In each of the groups additional work can be invested to improve the applicability of immobilized biocatalysts. The largest number of open questions exists with respect to the effects of kinetic stabilization and of microenvironmentel effects. The physiology of the microbial cells will play a key role in both cases and will have to be a primary concern in the future work of cell immobilization.

References

1.   I. Chibata: Immobilized Enzymes. Halsted Press, New York, 1978

2.   Enzymes and Immobilized Cells in Biotechnology, A.I. Laskin, ed. Benjamin Cummings Publ., 1985

3.   Appl. Biochemistry and Bioengineering, Vol. 4, L.B. Wingard, I. Chibata, eds., Academic Press, 1983

4.   J. Klein and F. Wagner, in Ref.3, p. 11-51

5.   J. Klein and K.-D. Vorlop: In: Comprehensive Biotechnology vol. 1, M. Moo-Young, ed., Pergamon Press, Oxford 1986, p. 203-224

6.   K.-D. Vorlop and J. Klein, in: Enzyme Technology, G. Lafferty, ed., Springer-Verlag, Heidelberg, 1983, 21-235

7.   J. Klein and F. Wagner, in: Dechema Monograph Vol. 82, 1978, 142-164

8.   K.-D. Vorlop, H.J. Steinert, J. Klein: Enzyme Engineering 8 (in press)

9.   J. Klein, J. Stock and K.-D. Vorlop: J. Appl. Microbial Biot. 18 , 1983, 86-91

10. J. Klein and B. Krebdorf, Biot. Letters 5 (1983) 497-502

11. J. Klein and B. Krebdorf, Biot. Letters 8 (1986)

12. J. Klein and F. Wagner, Dechema Monogr. Vo. 84, 1979, 265-335

13. J. Klein and B. Krebdorf, Biot. Letters 4 , (1982), 375-380

14. J. Klein, F. Wagner, B. Krebdorf, R. Muller, H. Tjokrosoeharto, K.D. Vorlop, in Ref. 2, p. 71-92

15. J. Klein and K.D. Vorlop, A.C.S. Sympos. Series Vol. 207, 1983, 377-392

16. J. Klein and P. Schara, Apll. Biochem. Biotechnol. 6 , 1981, 91-117

17. J. Klein and K.D. Vorlop, Chem. Eng. 7 , 1984, 233-240

18. J. Klein, K.D. Vorlop and F. Wagner, Enzyme Engineering 7 , Vol. 34, Ann. N.Y. Acad.Sci., 1984, 437-449

19. J. Klein and H. Eng, Biotechnol. Letters 1 , 1979, 171-174

20. J. Klein and M. Kluge, Biotechnol. Letters 3 , 1981, 65-70

21. K. Buchholz, ed., Dechema Monograph Vol. 84, Verlag Chemie 1979

22. E.F.B. Working Party: Enzyme Microb. Technol. 5 , 1983, 304-308

23. J. Klein and F. Wagner, Enzyme Engineering 8 , K. Mosbach ed. (in press)

24. J. Klein, U. Hackel and F. Wagner, ACS Symposium Ser. Vol. 106, 1979, 101-118

25. J. Klein and B. Krebdorf, Biot. Letters 4 , 1982, 375-380

26. P.M. Doran and J.E. Bailey, Biot. Bioeng. 28 ,1986, 1816-1831

27. J. Klein, J.W. Becke and K.D. Vorlop, Final Report of BEP of EEC, E. Magnien, ed. M. Nijhoff Publ. 1986, p. 181-194

28. S. Fukui and A. Tanabe, Enzyme Engineering 6 , 1982, 191-200

29. H.J. Steinert, Ph.D. Dissertation, Technical University Braunschweig, 1987

30. H.J. Steinert, K.D. Vorlop, J. Klein, 2. Symposium on Biocatalysis in Organic solvents, Waageningen, 1986

STRUCTURED MODELING OF IMMOBILIZED CELL KINETICS AND RNA CONTENT

HAROLD G. MONBOUQUETTE and DAVID F. OLLIS
Department of Chemical Engineering

North Carolina State University

Raleigh, North Carolina   27695-7905

ABSTRACT

A simple intrinsic structured model has been formulated to simulate steady-state substrate diffusion and consumption in porous, cell-laden carriers and to predict the RNA content of immobilized living (metabolically active) microorganisms. In addition, the model explicitly describes phenomena observed experimentally such as the thickness of a metabolically active cell layer and leakage of biomass from a support at a rate commensurate with immobilized-microbial growth. The technique of scanning microfluorimetry is being developed to determine experimentally the macromolecular composition of immobilized cells for comparison to structured model predictions.

INTRODUCTION

Unusual physiological states of metabolically active immobilized cells have been speculated upon widely as explanations for apparently anomolous kinetics of immobilized-cell fermentations. Porous carriers exert a wide variety of chemical and physical influences on immobilized microbes through, for example, unequal

9

partitioning of nutrients, products and protons between the bulk and the support microenvironment;[1-10] cell-carrier toxicity;[11] reduced water activity;[12] and mechanical resistance to cell growth.[13,14] These factors are presumed responsible for observed aberrant behavior including significantly increased or decreased generation times,[6,15,16] increased specific productivities,[15,17] increased product yields[17-19] and increased photosynthetic activity of immobilized alga.[20] A current lack of data concerning intra-biocatalyst mass transport complicates an accurate assessment of actual immobilized-cell physiological state since the intrinsic biocatalyst kinetics, which are a reflection of cell physiology, cannot be isolated easily from possible diffusional masking effects.[21,22]

A simple intrinsic structured model of broad applicability has been constructed as a qualitative, experimentally verifiable description of the diffusion and reaction phenomena and the physiological state of immobilized microbes under steady-state conditions.[23] The structured, two-compartment description[24,25] of immobilized biomass incorporated into the model facilitates prediction of intracellular RNA concentration with depth in porous carriers. Since cell RNA content correlates well to protein-synthesizing capability,[26] it is a useful indicator of cell physiological condition and intrinsic biocatalytic activity.

Novel application of the technique of scanning microfluorimetry is proposed as a means for model verification through determination of immobilized-cell macromolecular composition with depth in microscope-slide-mounted cross sections of cell-laden gel carriers. This powerful technique has long enjoyed widespread application in the field of histochemistry for measurement of the intracellular concentration of specific macromolecules.[27] Fluorochromes exist for the selective labeling of DNA, RNA, protein, carbohydrate and lipid.[28] Unlike flow cytometry,[29,30] the immobilized microbes do not have to be removed from the carrier into suspension before examination; the biocatalyst structure is preserved in cross section and the cells may be examined *in situ*. Measured values of immobilized-cell macromolecular composition will aid in the identification of immobilized-cell physiological states.

Recently, *S. cerevisiae* cells adsorbed in a monolayer to gelatin-coated glass beads were found to contain, using flow cytometry, an abnormally high amount of DNA yet a relatively low quantity of double-stranded RNA.[17] These results are in qualitative agreement with an earlier chemical analysis of average cellular DNA and RNA levels for polyacrylamide-gel-entrapped yeast.[31] The scanning microfluorimetric experimental approach would complement these studies nicely, as the *variation* in cell composition, hence physiology, could be investigated across a broad *cell layer in situ.*

THE STRUCTURED MODEL

The preponderance of published experimental evidence indicates that given a continuous supply of growth nutrients and an adequate inhibitory-product removal rate, immobilized cells grow to fill all accessible pore space within a carrier to a depth dictated by nutrient and/or inhibitory-product diffusional limitations.[15,16,32-34] When the pore volume fraction able to accomodate cells within this region is full, biomass leaks from the biocatalyst at a rate commensurate with continued *in situ* cell growth.[34-38] As a result, steady state is characterized by constant levels of product, residual nutrients and freely suspended biomass in the continuous-fermenter effluent.[15,39-46]

Cross sections of steady-state gel biocatalysts examined under the microscope typically reveal the features represented in Figure 1. A clearly defined, metabolically active surface/subsurface cell layer of nearly uniform density is often apparent.[15,39,40,47-53] The immobilized microbes colonize the carrier macropores; microporous regions inaccessible to biomass may also be evident within the cell layer.[14] The biocatalyst system, therefore, consists of three distinct phases: the biomass or *biotic phase*; the pore-filling fluid medium, the *abiotic phase*; and the

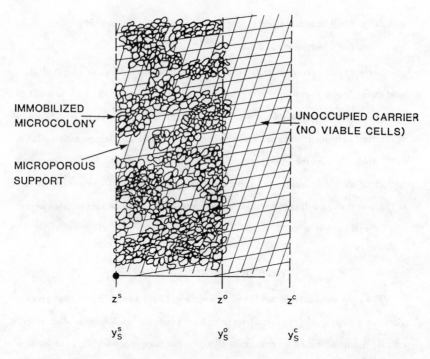

Figure 1.   Cross section of a slab-geometry, immobilized-cell biocatalyst. Adapted from reference 23.

carrier material itself.

Structured models, which are characterized by multi-variable descriptions of biomass, are particularly well-suited for application to these immobilized-cell systems since they by nature are intrinsic and thereby automatically provide for distinction between biotic and abiotic phases.[54] That is, cell-compartment concentrations must be expressed on a biotic-volume basis; whereas extracellular substrate and product concentrations logically should be based on the surrounding abiotic-phase volume. It follows that intra-biocatalyst transport of nutrients and products is limited to unoccupied micropores and any pore space between cells in immobilized microcolonies.

13

Esener *et al.*[24] have constructed and tested a two-compartment structured model. The hypothetical cell compartments divide cell dry weight into a synthetic component, R, composed primarily of ribosomal RNA (rRNA), and a structural/genetic compartment, D, mostly made up of protein and DNA. This model was found to simulate *K. pneumoniae*[24] and *Z. mobilis*[25] chemostat fermentations better than representative unstructured models.

A slightly modified form of the model developed by Esener *et al.*[24] is presented, with rate expressions, in block diagram form as Figure 2. The original model did not include the parameter, σ, an adjustment to the D-compartment synthesis term to account for the significant fraction of inactive rRNA at low specific

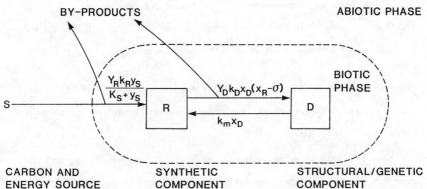

BY-PRODUCTS   **ABIOTIC PHASE**

BIOTIC PHASE

S

$\frac{Y_R k_R y_S}{K_S + y_S}$ R $Y_D k_D x_D (x_R - \sigma)$ D

$k_m x_D$

CARBON AND ENERGY SOURCE   SYNTHETIC COMPONENT   STRUCTURAL/GENETIC COMPONENT

Figure 2.   Block diagram of the two-compartment model proposed by Esener *et al.*[24] with modified rate expressions. Adapted from reference 23.

growth rates. With this minor adjustment, the model still simulates impressively the *K. pneumoniae* fermentation[24] while also accurately predicting the approximately linear increase in specific growth rate, μ, with RNA content.

As illustrated in Figure 3,[55] it is an established experimental result for a number of other bacterial species that specific growth rate increases almost linearly with steady-state intracellular RNA concentration.[26,55] Thus, model simulations

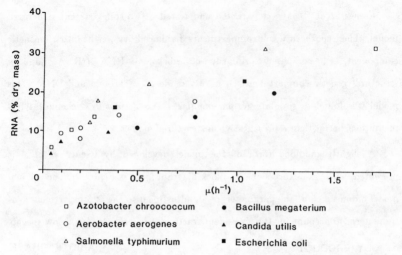

Figure 3.   RNA in percent dry weight versus specific growth rate for six
species of bacteria. Adapted from reference 55.

of immobilized-cell RNA content with depth into a carrier provide useful, experimentally verifiable predictions of intra-biocatalyst gradients in protein-
synthesizing activity, hence growth.

Since the biotic-phase concentrations of R and D compartment, x_R and x_D,
must sum to ρ_b, the assumed constant biotic-phase density (in units of dry weight
per unit wet biotic volume), the following steady state R-component balance fully
describes the biotic phase:

$$0 = \frac{Y_R k_R y_S}{K_S + y_S} - k_D(\rho_b - x_R)(x_R - \sigma) + k_m(\rho_b - x_R) - \mu x_R \tag{1}$$

where

$$\mu = \frac{1}{\rho_b}\left[\frac{Y_R k_R y_S}{K_S + y_S} + (Y_D - 1)k_D(\rho_b - x_R)(x_R - \sigma)\right] \tag{2}$$

Substrate (y_S) consumption to form R component with yield, Y_R, less than one follows Michaelis-Menten kinetics with rate constant, k_R, and saturation constant,

15

K_S. D-compartment synthesis follows second-order kinetics with imperfect yield, Y_D, and rate constant, k_D. D-component degradation, rationalized biologically as the turnover of macromolecules, is modeled as a first-order process with yield equal to one and rate constant, k_m. The final term on the right-hand side of equation 1 accounts for dilution of biomass from the carrier due to cell growth. Finally, equation 2 is obtained simply by summing all the biotic-phase conversions and dividing by ρ_b, generating an expression for the specific rate of biomass production, μ.[54]

The abiotic-phase equation which describes steady-state diffusion and consumption of a single limiting nutrient in a rectangular-geometry biocatalyst takes the familiar form,

$$D_e \frac{d^2 y_S(z)}{dz^2} = \left(\frac{\beta C_X}{\rho_b}\right) \frac{k_R y_S(z)}{K_S + y_S(z)} \tag{3}$$

where D_e is the effective diffusivity, z represents distance from the biocatalyst surface, β is the biocatalyst porosity, and C_X is the immobilized-biomass concentration based on the *total pore volume within the cell layer*. This concentration is assumed constant with position since, as discussed earlier, carrier cell loading is governed primarily by accessible pore volume fraction and internal diffusional limitations. A biomass concentration based on the *total biocatalyst volume delineated by the cell layer* may be substituted for the quantity, βC_X. In any case, the basis for the biomass concentration must be carefully derived and should be stated explicitly. The term, $\beta C_X/\rho_b$, is employed to convert the substrate uptake expression, $k_R y_S(z)/[K_S + y_S(z)]$, from a biotic-volume basis to a total-biocatalyst-volume basis. Note that the quantity k_R/ρ_b is equivalent to the rate constant for substrate uptake in unstructured models.

In order to explore, on a qualitative level, the effects of different cell loadings on biocatalyst performance, the effective diffusivity may be expressed as a linear function of the interstitial diffusivities in immobilized microcolonies, D_m, and unoccupied micropore space, D_c,

$$D_e = \beta[\epsilon\xi D_m + (1-\xi)D_c] \tag{4}$$

Here, ξ represents the accessible pore volume fraction and ϵ is the immobilized-microcolony porosity. The microcolony interstitial diffusivity may be significantly less than D_c due to an increase in diffusion path tortuosity and/or the presence of polymeric cell exudates.

Since immobilized-biomass concentration is assumed constant at steady state, continued immobilized-cell growth must correspond exactly to the biomass flux from the carrier. This cell leakage rate per unit biocatalyst surface area, q_b^s, may therefore be expressed as

$$q_b^s = \beta\, C_X \int_{z^c}^{z^s} \mu(z')dz' \tag{5}$$

with limits of integration at the center, z^c, and surface z^s, of the biocatalyst.

MODEL SOLUTION

Equation 3 may be integrated once analytically and subsequently solved numerically for the substrate profile with a fourth-order Runge-Kutta algorithm. The following boundary conditions are applied:

$$y_S(z^c) = y_S^c \qquad \frac{dy_S(z^c)}{dz} = 0$$

$$z = z^c \qquad y_S(z^s) = y_S^s$$

Note that under diffusion-limited conditions, the no-flux boundary condition becomes

$$y_S(z^c) = y_S^o \qquad \frac{dy_S(z^c)}{dz} = 0$$

In this case, if the minimum limiting-substrate concentration supportive of life, y_S^o, is known, the solution process is much simpler; otherwise the value for y_S^c must be estimated initially and eventually determined through an iterative process. Once the abiotic-phase substrate concentrations are known at each depth, R-compartment concentrations, specific growth rate and biomass flux outward at each point may be calculated sequentially in that order using equations 1, 2, and 5.

Figure 4 is a plot of y_S, μ and R-fraction (x_R/ρ_b) versus z for the parameter values listed in Table 1. The model predicts a metabolically active cell layer of about 220 μm in thickness. Substrate concentration, plotted semi-logarithmically, falls off rapidly near the surface and eventually reaches a minimum value, y_S^o $(=1.4 \times 10^{-3})$ g/l, corresponding to $\mu=0$ at the inner edge of the cell layer. This value for the minimum limiting-substrate concentration supportive of life, y_S^o, can be calculated *a priori* from equation 2 with μ set equal to zero and using a value for x_R^o from the y-intercept of a plot of x_R versus μ.

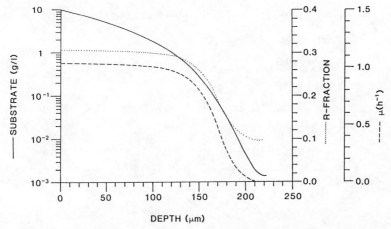

Figure 4.   Substrate, R-fraction and specific growth rate profiles with depth into the biocatalyst; $\eta_{CL}=0.75$. Parameter values: $y_S^s=10.0$ g/l, $\xi=0.90$, and $l_{BC}>l_{CL}$. From reference 23.

Due to the saturation-type kinetics for substrate uptake, the R-fraction and μ curves appear almost flat over a significant portion of the cell layer nearest the biocatalyst surface. A cell-layer effectiveness, based on immobilized-microbial growth may be defined to quantify this effect

TABLE 1[*]

Values for Model Parameters

Biotic Phase Equation	Abiotic Phase Equation
From Reference 24:	
$k_R = 470.4$ g/l·hr	$D_m = 3.5 \times 10^{-6}$ cm^2/sec
$k_m = 0.06$ hr^{-1}	$D_c = 7.0 \times 10^{-6}$ cm^2/sec
$Y_R = 0.73$	$\epsilon = 0.20$
$Y_D = 0.66$	$C_X/\rho_b = 0.72$
$K_S = 0.07$ g/l	$\xi = 0.90$
	$y_S^s = 10.0$ g/l
$x_R^o = 23.0$ g/l	$y_S^c = 1.393 \times 10^{-3}$ g/l
Estimated:	
$\rho_b = 240.0$ g/l	
Calculated:	
$k_D = 0.0311$ l/g·hr	
$\sigma = 20.1$ g/l	
$y_S^o = 1.393 \times 10^{-3}$ g/l	

[*] Adapted from reference 23

$$\eta_{CL} = \frac{q_b^s/\beta}{\mu^s C_X l_{CL}} \qquad (6)$$

where l_{CL} is the thickness of the cell layer. The denominator in equation 6 describes a rectangle on the μ versus z plot bounded by $\mu = 0$ and $\mu = \mu^s$ and $z^s = 0$ and $z^o = l_{CL}$. The numerator, q_b^s/β, is represented by the area under the curve.

For the case illustrated by Figure 4, $\eta_{CL} = 0.75$. Thus, even though the surface substrate concentration, 10 g/l, is much greater than K_S (= 0.07 g/l) a zero-order-kinetics modeling assumption under these conditions is a poor one. By raising y_S^s to 100.0 g/l, η_{CL} increases to 0.90, which is still significantly less than one; and the zero-order-kinetics assumption is still a poor assumption! The obvious route to improved effectiveness is to reduce biocatalyst size such that substrate concentration at the cell carrier center does not differ much from the surface concentration or such that y_S^c is still much greater than K_S at z^c.

Biocatalyst performance could also be expected to be sensitive to carrier pore size distribution, a manipulable quantity for many solid-support types, through its impact on the accessible-pore-volume fraction. A change in this quantity affects C_X directly since immobilized cells can grow to fill only the support pores or cavities large enough to accomodate them. A change in ξ thereby influences both effective diffusivity and substrate uptake (see equations 3 and 4).

As illustrated in Figure 5, the biomass flux, q_b^s, from an immobilized-cell biocatalyst, passes through a maximum with respect to ξ. The position and importance of this maximum depends on the relative values for the substrate interstitial diffusivities in carrier micropores and microcolonies. As the resistance to mass transfer in the microcolonies relative to that in unoccupied carrier increases, the sensitivity of q_b^s to ξ increases as well, yet the maximum q_b^s declines and is attained at lower values for ξ. Therefore, when D_m is much less than D_c, a carrier with a significant pore-volume fraction inaccessible to cells may be optimal, at least for growth-related product synthesis.

BIOCATALYST EFFECTIVENESS

A change in ξ, holding other parameters in Table 1 constant, results in alterations of the predicted cell layer thickness, but the relative shapes of the y_S, μ and R-fraction profiles remain unchanged as does η_{CL}. Thus, in order to adequately assess overall biocatalyst performance, a biocatalyst effectiveness, η_{BC}, based on substrate uptake, must be derived from equation 3 akin to that developed for inan-

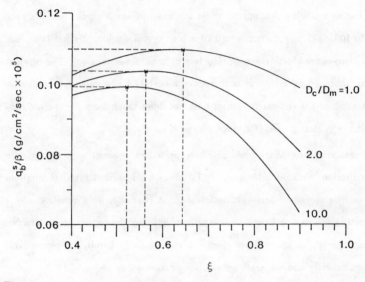

Figure 5.   Biomass production and accessible pore volume fraction; X denotes the maximum with respect to biomass leakage. Parameter values: $y_S^s = 10.0$ g/l, $D_c = 7.0 \times 10^{-6}$ cm²/sec, $l_{BC} > l_{CL}$. Adapted from reference 23.

imate heterogeneous catalysts,[56]

$$\eta_{BC} = \frac{\left\{2\left[\gamma^s - \gamma^c + \ln\left(\frac{1+\gamma^c}{1+\gamma^s}\right)\right]\right\}^{\frac{1}{2}}}{\phi\frac{\gamma^s}{1+\gamma^s}} \tag{7}$$

where $\gamma \equiv y_S/K_S$ and $\phi \equiv l_{BC}[\beta C_X k_R/(\rho_b D_e K_S)]^{\frac{1}{2}}$. When $\gamma^c = \gamma^o$, the kinetics within the biocatalyst are diffusion limited. Under these conditions, η_{BC} approaches the reciprocal of the generalized modulus, Φ.[56] Thus,

$$\Phi = \frac{\phi \dfrac{\gamma^s}{1+\gamma^s}}{\left\{ 2\left[\gamma^s - \gamma^o + \ln\left(\dfrac{1+\gamma^o}{1+\gamma^s} \right) \right] \right\}^{\frac{1}{2}}} \tag{8}$$

Note that Φ may be calculated analytically. Values for Φ may subsequently be used to estimate η_{BC} graphically from numerically generated plots of η_{BC} versus Φ.

Figure 6 indicates that for increasing γ^s the η_{BC} versus Φ curves approach the effectiveness-factor plot for a simple zero-order reaction system. However, this observation should not be misconstrued as evidence for approximately zero-order kinetics of immobilized-cell biocatalysts. This generalized modulus is not a zero-order modulus and even at high substrate concentrations the zero-order approximation may be a poor one as evidenced by the earlier η_{CL} calculations.

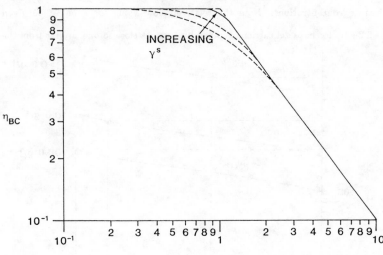

Figure 6.   Biocatalyst effectiveness based on substrate consumption versus the generalized modulus. The solid curve corresponds to $\gamma^s = 142.9$. The lower dashed curve refers to very small γ^s whereas the top broken curve corresponds to γ^s equal to infinity. Adapted from reference 23.

Semi-logarithmic plots of Φ versus ξ for three different surface substrate con-centrations, a biocatalyst thickness of 400 μm ($l_{BC} = 200$ μm) and Table 1 values for other parameters appear as nearly parallel curves of positive slope (see Figure 7). Clearly, the generalized modulus is quite sensitive to changes in ξ. Optimiza-tion of the accessible pore volume fraction with respect to the total substrate flux at the biocatalyst surface requires maximization of the following expression:

$$\eta_{BC}\left(\frac{\beta C_X}{\rho_b}\right)\frac{k_R y_S^s}{K_S + y_S^s}$$

Since $\xi = C_X/[\rho_b(1-\epsilon)]$, maximization of the above amounts to finding the largest value of the product of η_{BC} and ξ for given values of the remaining parameters noting the Φ, hence η_{BC}, is a function of ξ as well. The optimum values for ξ with respect to substrate uptake are indicated on Figure 7 for three different surface substrate concentrations. Note the relatively low value for ξ of ≈ 0.55 given $y_S^s = 0.2$ g/l. Perhaps gel carriers, typified by ξ values close to one, are far from the

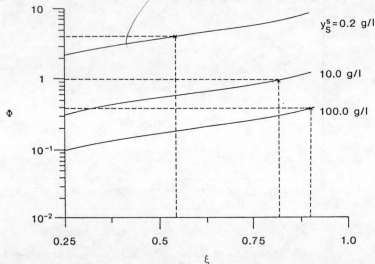

Figure 7.   Effect of accessible pore volume fraction on the generalized modulus; X denotes the optimum with respect to substrate con-sumption; $l_{BC} = 200.0$ μm. Adapted from reference 23.

ideal for maximum conversion of substrate in continuous stirred tanks or at the tail-end of packed beds where substrate concentrations are typically low.

SCANNING MICROFLUORIMETRY

The major implications and predictions of this model await experimental verification. Application of scanning microfluorimetry techniques should lead to discernment of differences in physiological states of immobilized and freely suspended cells and cells immobilized at various depths in a carrier based on observed variations in cell macromolecular composition. A number of fluorescent staining methods have been developed for DNA, RNA, protein, carbohydrate and lipid[28] so that messy chemical methods can be avoided, except as possible controls, and cells may be examined *in situ* within biocatalyst cross sections. The data gathered as to the gradient in intracellular RNA concentration may be at least qualitatively compared to the predictions of the structured model described above using appropriate estimates for kinetic parameters.

Incident-light microfluorimeters (epifluorescence microphotometers) have become the most popular equipment for quantitative examination of microscope slide preparations (see Figure 8). In this arrangement the exciting light impinges upon the sample from the objective rather than a substage condenser as for transmitted light microfluorimetry. For both types of equipment, the fluorescent light is collected by the objective and the shorter, exciting wavelengths and other stray light are filtered from the longer wavelength light before reaching the eyepiece and photomultiplier. Thus, the fluorescent image is viewed in a dark background making the technique very sensitive. Use of epifluorescence equipment further ensures the existence of a sharply contrasted fluorescent image against the dark background, obviates the necessity of condenser focusing, allows intermittent observation and focusing of the otherwise transparent image in longer wavelength transmitted light using phase-contrast optics, and permits study of opaque speci-

mens.[57]

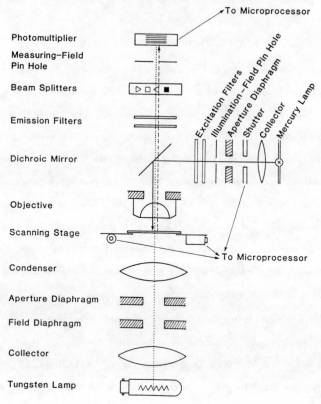

Figure 8.   Schematic of the OPELCO scanning microfluorimetry system consisting of Olympus and LEP equipment.

The addition of a microprocessor-controlled scanning stage, which facilitates precise, automated movement of a slide-mounted sample across the microscope viewing area, to the epifluorimeter enables scanning microfluorimetry. By coordinating stage movement with the photomultiplier measuring interval, fluorescence intensity at precise locations on a biocatalyst cross section (± 0.1 μm), for example, may be taken and stored for later processing on a minicomputer.

Our scanning microfluorimetry system, custom built with Olympus and LEP equipment by OPELCO of Washington, D.C., is fitted with additional features to facilitate useful information gathering (see Figure 8). Illumination field and measuring field pin-holes of variable size are incorporated into the optics to limit specimen photobleaching and to reduce the measuring area to the approximate size of a single cell, respectively. The microprocessor controlled shutter is employed to further limit exposure of the specimen to excitation light which often induces rapid fluorochrome fading. Finally, additional excitation and emission filters serve to further isolate exciting light from the optimum emitted light wavelengths for quantitative measurement.

For example, acridine orange, under properly controlled conditions, is specific for nucleic acids, moreover it stains RNA and DNA differently thereby allowing determination of RNA and DNA concentrations from a single preparation. The flat AO molecules intercalate between double-stranded DNA base pairs. Staining of primarily single-stranded RNA occurs through a cooperative binding process with negatively-charged phosphate groups.[58] With blue wavelength excitation, AO bound to DNA fluoresces green and AO-stained RNA fluoresces red.[28,29,30] Thus, concentrations of DNA and RNA can be determined from the intensity of light emitted at 515-580 nm and 610-650 nm, respectively.[58] DNase- and RNase-treated controls can be examined in parallel.[28]

CONCLUSIONS

An intrinsic model, which maintains the distinction between abiotic and biotic volumes in its description of effective diffusivity and limiting substrate uptake, qualitatively demonstrates the importance of the pore-volume fraction accessible to cells in biocatalyst design. When bulk substrate concentrations are low a ξ value approaching 0.5 may be optimal.

Incorporation of an appropriate two-compartment structural representation of biomass into the model facilitates prediction of intracellular RNA concentration with depth in a cell support. Since RNA content exhibits a near linear relationship with protein-synthesizing capability and specific growth rate, it is a useful measure of immobilized-cell physiological state.

The technique of scanning microfluorimetry offers a means to experimentally determine immobilized cell macromolecular composition, including RNA content, with position in the biocatalyst. Thus, the predictions of the structured model can be experimentally verified and the actual physiological state of immobilized cells can be established.

ACKNOWLEDGEMENT

This work was supported with a grant from the Celanese Corporation.

NOMENCLATURE

C_S substrate culture concentration (based on the total pore volume) $[M/L^3]$

C_X biomass (dry weight) culture concentration (based on the total pore volume within the metabolically active cell layer) $[M/L^3]$

D_c interstitial substrate diffusivity in carrier micropores $[L^2/T]$

D_e effective substrate diffusivity based on the total cross sectional surface area of the biocatalyst $[L^2/T]$

D_m interstitial substrate diffusivity in immobilized microcolony abiotic volume $[L^2/T]$

K_S Michaelis-Menten constant for substrate $[M/L^3]$

k_D rate constant for D-component formation $[L^3/(MT)]$

k_m rate constant for cell maintenance (D-component turnover) $[1/T]$

k_R rate constant for R-component formation (substrate uptake) $[M/(L^3T)]$

l_{BC} half the slab-biocatalyst thickness [L]

l_{CL} metabolically active cell-layer thickness [L]

q_b biomass (dry weight) flux $[M/(L^2 T)]$

x_D D-component biotic-phase concentration $[M/L^3]$

x_R R-component biotic-phase concentration $[M/L^3]$

Y_D yield of D component from R-component (dimensionless)

Y_R yield of R component from substrate (dimensionless)

y_S abiotic-phase substrate concentration $[M/L^3]$

z distance from the slab-biocatalyst surface [L]

Greek

β cell-carrier porosity (dimensionless)

ϵ abiotic-phase fraction of immobilized-microcolony volume (dimensionless)

γ dimensionless substrate concentration (y_S/K_S)

η_{BC} theoretical biocatalyst effectiveness (dimensionless)

η_{CL} theoretical metabolically active cell-layer effectiveness (dimensionless)

μ specific growth rate [1/T]

ξ pore-volume fraction accessible to cells (dimensionless)

ρ_b biomass density (dry weight/wet biotic volume) $[M/L^3]$

σ correction for "inactive" ribosomes $[M/L^3]$

Φ generalized modulus (dimensionless)

ϕ pseudo-first-order modulus (dimensionless)

Superscripts

c   value at the slab-biocatalyst center line

o   value when $\mu = 0$

s   value at the biocatalyst surface

REFERENCES

1. Abbot, B.J.: in: Ann. Rep. Ferm. Proc. (Perlman, D., ed.), Vol. 1, p. 205, New York: Academic Press, 1977

2. Martin, J.P., Filip, Z. and Haider, K.: Soil Biol. Biochem. *8*, 409 (1976)

3. King, V.A.-E. and Zall, R.R.: Proc. Biochem. *18*(6), 17 (1983)

4. Chibata, I., Tosa, T. and Takata, I.: Trends Biotech. *1*, 9 (1983)

5. Ehrhardt, H.M. and Rehm, H.J.: Appl. Microbiol. Biotech. *21*, 32 (1985)

6. Hattori, R.: J. Gen. Appl. Microbiol. *18*, 319 (1972)

7. Hattori, T. and Furusuka, C.: J. Biochem. *48*, 831 (1960)

8. Hattori, T. and Furusuka, C.: J. Biochem. *50*, 312 (1961)

9. Day, D.F. and Sarkar, D.: Enz. Eng. *6*, 343 (1982)

10. Williams, D. and Munnecke, D.M.: Biotech. Bioeng. *23*, 1813 (1981)

11. Tramper, J., Suwinska-Borowiec, G. and Klapwijk, A.: Enz. Microb. Technol. *7*(4), 155 (1985)

12. Mattiasson, B. and Hahn-Hägerdal, B.: Eur. J. Appl. Microbiol. Biotech. *16*, 52 (1982)

13. Somerville, H.J., Mason, J.R. and Ruffell, R.N.: Eur. J. Appl. Microbiol. Biotech. *4*, 75 (1977)

14. Inloes, D.S., Smith, W.J., Taylor, D.P., Cohen, S.N., Michaels, A.S. and Robertson, C.R.: Biotech. Bioeng. *25*, 2653 (1983)

15. Shinmyo, A., Kimura, H. and Okada, H.: Eur. J. Appl. Microbiol. Biotech. *14*, 7 (1982)

16. Baudet, C., Barbotin, J.-N. and Guespin-Michael, J.: Appl. Environ. Microbiol. *45*, 297 (1983)

17. Doran, P.M. and Bailey, J.E.: Biotech. Bioeng. *28*, 73 (1986)

18. Navarro, J.M. and Durand, G.: Eur. J. Appl. Microbiol. Biotech. *4*, 243 (1977)

19. Holeberg, I.B., and Margalith, P.: Eur. J. Appl. Microbiol. Biotech. *13*, 133 (1981)

20. Bailliez, C., Largeau, C., Berkaloff, C., Casadevall, E.: Appl. Microbiol. Biotech. *23*, 361 (1986)

21. Ollis, D.F.: Biotech. Bioeng. *24*, 871 (1972)

22. Ooshima, H. and Harano, Y.: Biotech. Bioeng. *23*, 1991 (1981)

23. Monbouquette, H.G. and Ollis, D.F.: Ann. N.Y. Acad. Sci. In press (1986)

24. Esener, A.A., Veerman, T., Roels, J.A. and Kossen, N.W.F.: Biotech. Bioeng. *24*, 1749 (1982)

25. Jöbses, I.M.L., Egberts, G.T.C., van Baalen, A. and Roels, J.A.: Biotech. Bioeng. *27*, 984 (1985)

26. Ingraham, J.L., Maaløe, O. and Neidhardt, F.C.: Growth of the Bacterial Cell, Sunderland, MA: Sinauer, 1983

27. Swift, H.: in: Introduction to Quantitative Cytochemistry (Wied, G.L., ed.), pp. 1-39, New York: Academic Press, 1966

28. Kasten, F.H.: in: Staining Procedures (Clark, G., ed.), pp. 39-103, Baltimore: Williams and Wilkins, 1981

29. Kruth, H.S.: Anal. Biochem. *125*, 225 (1982)

30. Muirhead, K.A. et al. Bio/tech. *3*, 337 (1985)

31. Siess, M.H. and Divies, C.: Eur. J. Appl. Microbiol. Biotech. *12*, 10 (1981)

32. Garde, V.L., Thomasset, B. and Barbotin, J.-N.: Enz. Microb. Technol. *3*, 216 (1981)

30

33. Dhulster, P., Barbotin, J.-N., and Thomas, D.: Appl. Microbiol. Biotech. *20*, 87 (1984)

34. de Taxis du Poët, P., Dhulster, P., Barbotin, J.-N., and Thomas, D.: J. Bac. *165*(3), 871 (1986)

35. Burrill, H.N., Bell, L.E., Greenfield, P.F. and Do, D.D.: Appl. Env. Microbiol. *46*, 716 (1983)

36. Wada, M., Kato, J. and Chibata, I.: J. Ferm. Technol. *58*, 327 (1980)

37. Kuek, C. and Armitage, T.M.: Enz. Microb. Technol. *7*, 121 (1985)

38. Bailliez, C., Largeau, C. and Casadevall, E.: Appl. Microbiol. Biotechnol. *23*, 99 (1985)

39. Wada, M., Kato, J. and Chibata, I.: Eur. J. Appl. Microbiol. Biotech. *10*, 275 (1980)

40. Osuga, J., Mori, A., and Kato, J.: J. Ferm. Technol. *62*, 139 (1984)

41. Wada, M., Uchida, T., Kato, J. and Chibata, I.: Biotech. Bioeng. *22*, 1175 (1980)

42. Krouwel, P.G., Groot, W.J., Kossen, N.W.F. and van der Laan, W.F.M.: Enz. Microb. Technol. *5*, 46 (1983)

43. Messing, R.A. and Stineman, T.L.: Ann. N.Y. Acad. Sci. *413*, 501 (1983)

44. Margaritas, A and Bajpai, P.: Ann. N.Y. Acad. Sci. *413*, 479 (1983)

45. Klein, J. and Kressdorf, B.: Biotech. Lett. *5*, 497 (1983)

46. Prevost, H., Divies, C. and Rousseau, E.: Biotech. Lett. *7*(4), 247 (1985)

47. Nagashima, M., Azuma, M., Noguchi, S., Inuzuka, K. and Samejima, H.: Biotech. Bioeng. *26*, 992 (1982)

48. Eikmeier, H., Westmeier, F. and Rehm, H.J.: Appl. Microbiol. Biotech. *19*, 53 (1984)

49. Kopp, B. and Rehm, H.J.: Appl. Microbiol. Biotech. *19*, 141 (1984)

50. Bettman, H. and Rehm, H.J.: Appl. Microbiol. Biotech. *20*(5), 285 (1984)

51. Robinson, P.K. and Dainty, A.L.: Enz. Microb. Technol. *7*(5), 212 (1985)

52. Gosman, B. and Rehm, H.J.: Appl. Microbiol. Biotech. *23*, 163 (1986)

53. Mahmoud, W. and Rehm, H.J.: Appl. Microbiol. Biotech. *23*, 305 (1986)

54. Fredrickson, A.G.: Biotech. Bioeng. *18*, 1481 (1976)

55. Harder, A. and Roels, J.A.: in: Adv. Biochem. Engin. (Fiechter, A., ed.), Vol. 21 p. 55, Berlin: Springer-Verlag, 1982

56. Froment, G.F. and Bischoff, K.B.: Chemical Reactor Analysis and Design, pp. 182-184, New York: John Wiley and Sons, 1979

57. Ruch, F.: in: Introduction to Quantitative Cytochemistry II (Wied, G.L. and Bahr, G.F., eds.), pp. 431-450, New York: Academic Press, 1970

58. West, S.S.: in: Physical Techniques in Biological Research, Vol. III, Part C (Pollister, A.W., ed.), pp. 253-321, New York: Academic Press, 1966

ANALYSIS OF OXYGEN TRANSPORT IN IMMOBILIZED WHOLE CELLS

Ho Nam Chang* and Murray Moo-Young
Department of Chemical Engineering
University of Waterloo
Waterloo, Ontario, Canada N2L 3G1

* On sabbatical leave from KAIST, Seoul, Korea

ABSTRACT

Oxygen transport routes to immobilized cells in various bioreactors were identified in terms of mass transfer resistances, and oxygen uptake rates of immobilized cells in fluidized beds were calculated by taking into account maximum oxygen demand rate, gas holdup, solid holdup, superficial gas velocity and bead size. Oxygen penetration depths were influenced most by oxygen demand rate and diffusion coefficient of oxygen in immobilized cells. Penetration depths of oxygen in microbial cell systems are around 50-200 μ and those of animal and plant tissue cultures are around 500 - 1000 μ.

INTRODUCTION

Adequate aeration in aerobic fermentations is important. In bubble columns and air-lift fermenters, pneumatic energy promotes oxygen supply and mixing while mechanical stirring plays a greater role in stirred tank reactors, which for applications to immobilized whole cell systems, the former has received more attention than the latter because of the fragile nature of the immobilized system. Bubble columns or other pneumatically-agitated types of reactors can provide more oxygen transfer rate than stirred tank reactors in immobilized systems. For example, Gbewonyo and Wang (1) improved the oxygen supply rate and penicillin productivity using *Penicillium chrysogenum* adsorbed to microporous celite beads. Oxygen transfer in the bubble column was facilitated apparently by reduced broth viscosity and as a result more penicillin G was produced. Chang et al. (2) showed that more gluconic acid was produced with glucose oxidase immobilized in a dual hollow fiber reactor than with glucose oxidase immobilized in a Ca-alginate beads in a rotating packed disk reactor (3). In the dual hollow fiber reactor, oxygen was directly transferred to the immobilized glucose oxidase through the silicone membrane of the reactor.

33

In general, immobilization of whole cells (microbial, animal or plant) creates an environment in which oxygen transfer to the cells become more difficult than in free cell suspension cultures where the oxygen transfer is often limited by a gas-liquid interface. Cells immobilized to solid supports tend to form dense aggregates. The oxygen transfer in immobilized cells differs from that in free cell suspension cultures in two aspects: the densely-packed cells demand more oxygen per unit volume and the supply of oxygen should cope with additional mass transfer resistances such as the liquid-solid interface and intraparticle diffusion. Essentially, we have three-phase oxygen mass transfer problems in contrast to the two-phase problems of free cell suspension cultures. In this study, we will analyze various oxygen transport steps in immobilized whole cell systems and identify a rate-limiting step in O_2 transport using a fluidized bed system as an example. The results should be useful in the design of immobilized whole cell reactors.

OXYGEN TRANSFER IN IMMOBILIZED WHOLE CELL REACTORS

Table 1 lists existing three-phase bioreactors types classified according to the location of the cell aggregates (4). Strictly speaking, category IV is not considered as immobilized cells, but it is included here since the purpose of employing such reactors are the same as that of true immobilized cells. Recently, Enzminger and Asenjo (5) used a membrane cell recycle reactor to study citric acid production by yeast. Simple hollow fiber reactors were used to immobilize animal cells by Knazek et al. (6) in the early 1970's and *E. coli* cells by Inoles et al. (7) in 1983. Since a simple hollow fiber system could not provide enough oxygen needed for microbial cell growth, a dual hollow fiber system employing silicone tubing for oxygen transfer was devised by Robertson and Kim (8). Chang et al. (9) produced rifamycin B successfully using a modified dual hollow fiber reactor for more than 50 days. Trickle bed reactors are now very much in use in the treatment of waste water. Briffaud and Engasser (10) used this reactor type to produce citric acid. Rotating packed bed reactor (RPDR) was used to produce gluconic acid with immobilized oxidase (3) and ethanol with immobilized yeast cells (11). One advantage of RPDR and RBC is the relative ease of compartmentalization such that the reactors perform closely like plug flow reactors. Fluidized bed reactors find most applications to immobilized cell systems because of the gentle mixing characteristics (12,13). Webb et al. (14) used a spouted bed to produce cellulase. Except for animal and plant cells that have relatively low oxygen demand, no reactors for immobilized cells requiring very high oxygen requirements are commercially used at present. Many reactor types are under consideration, including solid-state fermentation recently used for cellulase production (15).

TABLE 1:   Three-phase bioreactors for immobilized cells

I. Stationary Particle (Surface) Reactors

    1. Hollow Fiber Reactors
       - simple hollow fiber
       - dual hollow fiber

    2. Packed Bed
       - simple packed bed (liquid-phase continuous)
       - trickle packed bed (gas-phase continous)

    3. Petri Dish, Solid culture

II. Moving Surface Reactors

    1. Rotating Packed Disk Reactor
    2. Rotating Biological Contactor
    3. Roller Bottle

III. Mixed Particle Reactors

    1. Stirred Tank
    2. Fluidized Bed Reactor
    3. Spouted Bed
    4. Circulating Bed Reactor
    5. Tower Fermentor

IV. Free Cell Recycle Reactors

    1. Membrane Recycle - flat, hollow fiber
    2. Centrifuge
    3. Settling Device - activated sludge

    In the design of three-phase bioreactors, it is important to understand how oxygen is transferred from the gas-phase to cells confined in the supports. Enfors and Mattiasson (16) reviewed oxygenation processes involving immobilized cells and Radovich (17) emphasized the importance of mass transfer in immobilized cell systems. Oxygen can be transferred in a one-stage reactor where three phases (gas, liquid and solid) exist, or in a two-stage reactor where the oxygenation and reaction processes are separated (Figure 1). Oxygen existing in a continuous gas phase can be transferred directly to cells such as in solid culture or through silicone tubes to immobilized cells such as in the dual hollow fiber reactor types. Both cases do not involve a liquid-phase as a transport medium. In most reactors, oxygen in bubbles is transferred to the liquid and soluble oxygen is

I. __One-Stage Reactor__

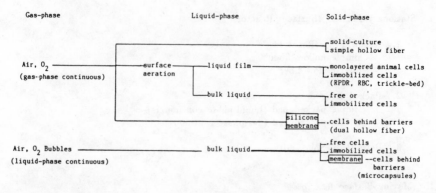

II. __Two-Stage Reactor__

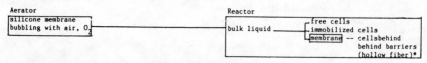

* O_2 transport through the lumen-side of the hollow fiber

Figure 1.   Oxygen Transfer Routes to Immobilized Cells

transported to the solid surface. Then, the oxygen diffuses through the solid support to reach the immobilized cells. In the case of microencapsulation, oxygen should go through one more barrier, the coating membrane of the microcapsules. In two-stage systems, aeration can be achieved by a silicone membrane (2) or bubbling with air or O_2 and liquid saturated with O_2 is supplied to the reactor. Since aeration is not involved in these reactor types, the design and operation becomes simpler compared to one-stage reactor systems.

Moo-Young and Blanch (18) listed eight possible oxygen transport steps to immobilized cells suspended in liquid solution. Among these, the mass transfer steps in a liquid film at the gas-liquid interface, in the bulk liquid, at the liquid-solid interface and in the solid-phase containing cells are likely to be important. If bulk mixing is good, the bulk liquid phase transport step can be neglected. As shown in Figure 1, different reactors have different O_2 transport routes. Figure 2 shows significant transport steps of oxygen in fluidized bed reactors, microcarrier and microcapsules, and dual hollow fiber systems. More transport steps do not necessarily mean more transport resistance. In the case of microcarriers and microcapsules, mass transport around the support may not be as important as that of the liquid film around the air bubbles because of the small support size. In example 3, the O_2 transport steps were reduced to 2 steps, but the glucose steps were increased to 3 steps instead of 2 steps in fluidized bed reactor types.

1. Fluidized Bed Reactor Systems

3 steps
- liquid film around bubble
- liquid film around solid
- inside the solid

2. Microcarrier and microcapsules

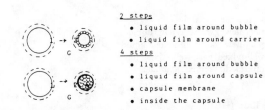

2 steps
- liquid film around bubble
- liquid film around carrier

4 steps
- liquid film around bubble
- liquid film around capsule
- capsule membrane
- inside the capsule

3. Dual Hollow Fiber Systems

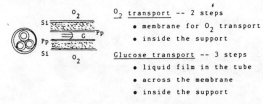

O_2 transport -- 2 steps
- membrane for O_2 transport
- inside the support

Glucose transport -- 3 steps
- liquid film in the tube
- across the membrane
- inside the support

Figure 2. Significant Oxygen Transport Steps in Immobilized Cell Systems

Oxygen transfer rates (OTR) and oxygen uptake rates (OUR) in fluidized beds can be influenced by a number of factors. Local oxygen transfer near the immobilized cells can be affected by a number of potential rate-limiting steps: maximum oxygen demand rate (ODR), gas holdup, solid holdup, superficial gas velocity and bead size. The overall oxygen requirement will also be determined by whether the gas and liquid phases are in mixed or plug flow modes. Since the extremes of plug-flow or well-mixed models can not describe O_2 transport in bubble columns properly, it is necessary to use intermediate mixing models in columns higher than 5 m. Oxygen uptake rate (OUR) can depend on the specific oxygen uptake of immobilized cells and their physiological state, and the number of cells per unit volume of support (16). Since the O_2 transport in the reactor scale is beyond the scope of this work, we will not deal with it here. The important factors for OTR and OUR are summarized in Table 2.

38

TABLE 2: Important variables in oxygen transfer rate and oxygen uptake rate

Oxygen Transfer Rate (OTR)

1. Microscale -- local oxygen transfer near immobilized cells
               - how many steps are involved?
               - what is rate-limiting step?
               - max oxygen demand rate (ODR), gas holdup, solid fraction, superficial
               gas velocity, bead size.

2. Macroscale -- reactor dynamics
               - plug-flow gas, plug-flow liquid
               - plug-flow gas, well-mixed liquid
               - well-mixed gas, well-mixed liquid
               - intermediate mixing models

Oxygen Uptake Rate (OUR)

1. Cellular level
               - specific oxygen uptake rate
               - physiological state

2. Number of Cells
               - unit volume of support
               - unit volume of reactor

CALCULATION OF OXYGEN TRANSFER RATE

Oxygen transfer from air bubbles to immobilized cells in bubble columns or air-lift fermentors can be represented by three steps (Figure 2, example 1). The rate equation for total oxygen transfer rate from gas to liquid is given by

$$r_b = k_L a_L \phi_L (C_b^* - C_b) \tag{1}$$

where $k_L a_L$ refers to O_2 transfer rate and ϕ_L is the volume fraction of liquid. Thus r_b refers to the rate per unit volume of the reactor rather than unit volume of liquid. The oxygen transfer from liquid to the solid is

$$r_f = k_s a_s \phi_s (C_b - C_{bi}) \tag{2}$$

where k_s refers to mass transfer coefficient for oxygen from liquid to the solid and ϕ_s is the solid particle fraction. C_{bi} is the interfacial oxygen concentration at the solid surface. Finally, the oxygen flux from the solid-surface to the interior is

$$r_s = -D_{eff} \frac{dC_s}{dr}\bigg|_{r=r_i} \cdot a_s \phi_s \tag{3}$$

where C_s is the oxygen concentration in the solid-phase. At steady-state the following relationship holds

$$r_b = r_f = r_s \tag{4}$$

$$\phi_g + \phi_L + \phi_s = 1 \tag{5}$$

Equations (1) and (2) are combined to give

$$r_b = r_f = k_{f,\,eff}\; a_s \phi_s (C_b^* - C_{bi}) \tag{6}$$

where

$$k_{f,\,eff} = \cfrac{\dfrac{1}{a_s \phi_s}}{\dfrac{1}{k_L a_L \phi_L} + \dfrac{1}{k_s a_s \phi_s}}$$

$$= \frac{k_L a_L \phi_L \bullet k_s}{k_L a_L \phi_L + k_s a_s \phi_s} \tag{7}$$

If gas-liquid phase mass transfer is rate-controlling ($k_L a_L \phi_L \ll k_s a_s \phi_s$)

$$k_{f,\,eff} = k_L \bullet \frac{a_L \phi_L}{a_s \phi_s} \tag{8}$$

On the other hand if liquid-solid phase mass transfer is rate controlling ($k_L a_L \phi_L \gg k_s a_s \phi_s$)

$$k_{f,\,eff} = k_s \tag{9}$$

If both resistances are important, equation (7) is used. The oxygen penetration depths can be calculated by the method given in the appendix and the details of the method for other geometries can be found elsewhere (19). Once the penetration depth is obtained, the total oxygen uptake rate (TOUR) can be obtained by the following equation.

$$TOUR = k_{f,\,eff}\; (C_b^* - C_{bi})\; a_s \phi_s \tag{10}$$

$k_L a_L$ values for the gas-liquid mass transfer rate were estimated by the correlation of Shah et al. (20).

$$k_L a_L = 0.467\; (0.01\; Usg)^{0.82} \tag{11}$$

where Usg is the superficial gas velocity given in cm/sec. k_s values were estimated by Deckwer's correlation (21).

$$k_s = D_L/dp\; (2 + 0.545(\nu_L/D_L)^{1/3}\; (\phi_g dp^4/\nu_L)^{0.214}) \tag{12}$$

where dp is the particle diameter and ν_L is the kinetic viscosity of liquid. D_L is the diffusion coefficient of oxygen in liquid.

RESULTS AND DISCUSSION

Effect of oxygen diffusivity and demand rate on penetration depth

The physical property most relevant to the estimation of oxygen penetration depth is the effective diffusivity of oxygen in the aggregate. The oxygen diffusivities were shown to vary from 10% to 110% of that in water (4). Recently, Wittler et al. (22) showed that the oxygen diffusivities in *Penicillium chrysogenum* pellets were increased as many as 8 times the molecular diffusivity by turbulent diffusion and convection. This large diffusivity may be due to the loose nature of the pellets and partly due to the experimental procedure which subjected the pellets perpendicularly to the liquid flow. In general, it is expected that the relative diffusivity of oxygen in the support is less than that in water. Figure 3 shows the penetration depth of oxygen in a spherical support of 2 mm in diameter as a function of relative diffusivity. When the oxygen demand rate Q was 100 mM $O_2/(L_s \cdot h)$ where L_s is the liter support volume, the penetration depth changed from 40 μ at 10% relative diffusivity to 80 μ at 100% relative diffusivity. The change becomes less sensitive as Q becomes higher.

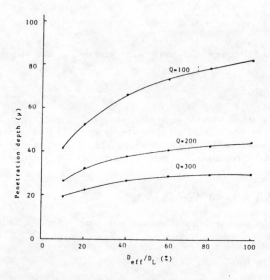

Figure 3. Oxygen Penetration Depth in a Spherical Support with Immobilized Cells

D_L=1.6 x 10^{-5} cm^2/s, U_{sg}=2 cm/s, ϕ_s=0.1, ν_L=0.01 cm^2/s, C_b=2.5 x 10^{-4} mM O_2

The most important single variable determining oxygen penetration depth is oxygen demand rate of immobilized cells. In Table 3 several model organisms from bacteria to plant tissue culture have been selected and maximum possible oxygen demand rates were calculated with specific uptake rate and maximum cell density. Most immobilized cells would require much less oxygen since they are more or less in the resting state. Thus the values reported in Table 3 represents the upper maxima of oxygen demand rates. In Figure 4 are shown penetration depths of oxygen for various organisms listed in Table 3 when they are immobilized in a 2 mm sphere, as a function of oxygen demand rate per unit volume of support. In the case of plant and animal tissue cultures, penetration depths are in the range of 400 - 1000 μ, but with *A. niger* and *P. chrysogenum* that have higher oxygen demands the penetration depths are in the range of 20 - 200 μ. Since *E. coli* and *S. cerevisiae* are facultative organisms, they will grow even under circumstances of low oxygen supply. In Figure 4, penetration depths of oxygen are shown when external diffusion limitation is removed. For high oxygen demanding cases, the gap is very wide, but it decreases with the decreasing Q. Huang and Bungay (30) measured oxygen penetration depths in an *A. niger* pellet. The pellet was 6 mm in diameter and 15 g dry weight of *A. niger*/liter of support. Q will be between 15 mM $O_2/L_s \cdot h$ and 45 mM $O_2/L_s \cdot h$ depending on the specific oxygen uptake rate. From Figure 4 this gives penetration depths between 110 μ and 300 μ; 150 μ is found. Thus, we can see that the current method is useful in estimating approximate oxygen penetration depths.

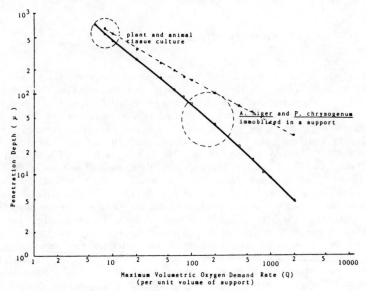

Figure 4. Oxygen Penetration Depths as a function of Q

d_p = 2 mm, D_{eff}/D_L = 0.7, the other conditions are the same as in Figure 3. The dotted line represent the penetration depth in the absence of external diffusion limitation.

TABLE 3: Estimated maximum oxygen uptake rates of representative organisms

	Specific OUR	Max. Cell Density	Max. OUR (mM O_2/L)
1. Bacteria			
Escherichia coli	10.8 mM O_2/(g.h) (26)	600 g/L$_s$ (9)	6480
2. Yeasts			
Saccharomyces cerevisiae	8.0 mM O_2/(g.h)	210 g/L (23)	1680
3. Filamentous Fungi		150 g/L$_s$(24)	1200
Aspergillus niger	3.0 mM O_2/(g.h)	100 g/L$_s$(e)	300
(citric acid)			
Nocardia mediterranei	3.0 mM O_2/(g.h)	550 g/L$_s$(9)	1650
4. Animal tissue culture			
Chick embryo fibroblast(25)	$(0.4-1.0) \times 10^{-10}$	5×10^9	0.5-50$_{(27)}$
5. Plant tissue culture	mM O_2/(cell.h)	cells/L	(7-10.)
Capasicum frutescens	6.72×10^{-3}(28)	1.07 g.fr.wt/	7-19
Digitalis lanta	mM O_2/(g.fr.wt.h)	cm^3 support (29)	

TOTAL OXYGEN UPTAKE RATE (TOUR)

Total oxygen uptake rate of immobilized cells is an indicator of the oxygen mass transfer efficiency in response to the oxygen demand of the immobilized cells. When the demand is low, the penetration depth reaches the center of the support meaning that no cells are starving of oxygen, but only a fraction of the cells near the surface of the support have enough oxygen for survival when the demand is high. Figure 5 (a) shows the total oxygen uptake rate calculated by equation (10). When the demand rate changes from 10 to 1000 by a factor of 100, TOUR changes from 0.7 mM/(L•h) to 2.8 mM/O_2/(L•h). Thus, under oxygen limiting cases the increase of biomass in the support by a factor of 2, say from 100g/L to 200 g/L, would result in only 15% increase in TOUR. In the same figure the ratio of TOUR without any external diffusion limitation to TOUR is shown. As was in the previous figure, the external diffusion limitation has a larger effect when Q is high. This is well represented in Figure 5b where the oxygen concentration at the support surface is very low when Q is high. If Q is above 400, then the surface oxygen concentration falls below the critical oxygen concentration which is usually 10% of the oxygen concentration saturated with air. In large support particles with dense cells, cells will starve of oxygen. As in the penicillin fermentation by Gbewonyo and Wang (1), it would be wise to immobilize high oxygen demanding cells on the surface of small particles to avoid oxygen limitation problems.

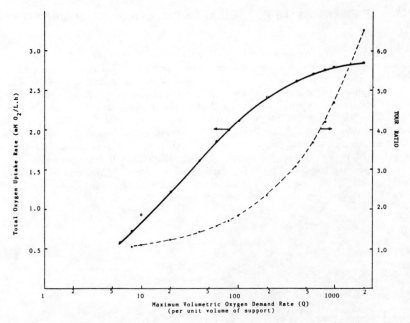

Figure 5
  (a) Total Oxygen Uptake Rate as a Function of Q. The conditions are the
      same as those in Figure 4. TOUR Ratio refers to the ratio of TOUR
      in the absence of external diffusion to TOUR.

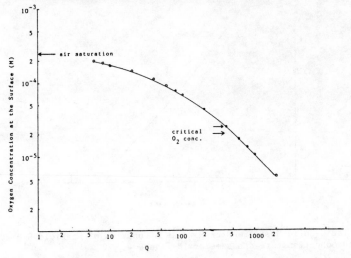

  (b) Oxygen Concentration at the Surface of an Immobilized Cell Particle.
      The conditions are the same as before.

MASS TRANSFER RATES IN THE GAS-LIQUID AND LIQUID-SOLID INTERFACES

Estimation of volumetric gas-liquid mass transfer coefficient has been of great importance to the design of aerobic fermentors and other gas-liquid contacting devices. For this reason many reliable correlations on gas holdup and k_La were developed for various applications (20). But there is no correlation yet available for three-phase reactor systems involving immobilized cells. In this study we used the simple correlation by Shah et al. (20) applicable to air water system. The presence of solid particles were not taken into account. The liquid-solid mass transfer coefficient k_s was estimated by Deckwer's correlation (21). k_s is an inverse function of d_p but proportional to the power of gas holdup, ϕ_g 0.264. Decreasing particle diameter will increase liquid-solid mass transfer rate as shown in Figure 6a. Superficial gas velocity has an effect both on the liquid-solid and gas-liquid mass transfer coefficients (Figure 6b). As superficial gas velocity increases, overall mass transfer rate increases.

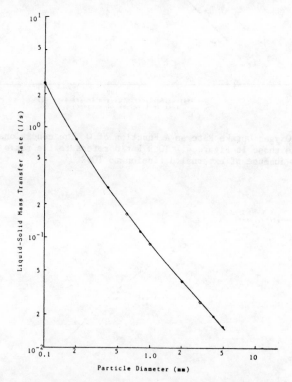

Figure 6
   (a) Liquid-Solid Mass Transfer Rate as a Function of Particle Diameter.
   The conditions are the same as in Figure 4.

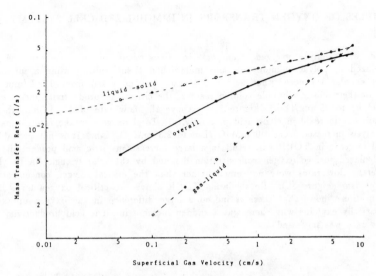

(b) Mass Transfer Rate versus Superficial Gas Velocity, U_{sg}

Since the reactor volume is constant, increasing solid content means a decreased space for gas-liquid mass transfer. Thus, at a higher solid fraction, overall mass transfer rates based on the reactor volume decrease (Figure 7). The total oxygen uptake increases when the solid faction is low and TOUR begins to decrease above a solid fraction of 0.4. Chang et al. (3) found that in their rotating packed disk reactor with immobilized oxidase increasing liquid level in the reactor has a similar effect. Thus in three-phase bioreactors there must be optimum levels of gas-holdup and solid content.

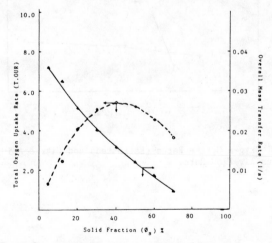

Figure 7. Total Oxygen Uptake Rate versus Solid Fraction of Particles

EXAMPLES OF OXYGEN TRANSPORT IN IMMOBILIZED CELL SYSTEMS

We will analyze oxygen transport in citric acid production using *A. niger* immobilized in polyacrylamide gels and mammalian tissue culture using a microcarrier. Horitsu et al. (31) noted that as air flow rate was increased from 0.7 L/min to 2.8 L/min in their 300 mL bubble column reactor, the citric acid productivity increased from 1 mg/(L.h) to 25 mg/(L.h) (Figure 8a). Above the flow rate of 1.4 L/min, there was no significant increase in citric acid productivity. Total oxygen uptake rate estimated by our analysis increased from 1.9 mM O_2/L•h to 2.3 mM O_2/L•h. It is not probable that a small increase in TOUR can result in a large increase in citric acid productivity. But the calculated surface oxygen concentration denoted by the solid triangle (in the Figure) at different flow rates were not much higher than the critical oxygen concentration for this organism (Figure 8b). Recalculating TOUR above the critical oxygen level (10% of air saturation) shows that there is indeed a large difference in the oxygen uptake rate. This partially explains why there was a sudden increase in citric acid productivity as the air flow rate was increased.

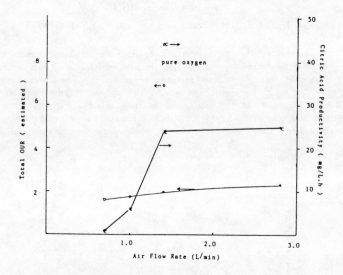

Figure 8
   (a) Total Oxygen Uptake Rates (Estimated) and Citric Acid Productivity versus Air Flow Rate.

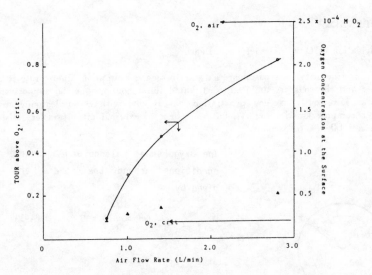

(b) Total Oxygen Uptake Rate above Critical Oxygen Concentration.
The Oxygen Concentrations at the Surfaces are given by ▲ .

Because of their large surface areas, microcarrier beads are used for animal tissue culture. Oxygen transfer in a microcarrier system can be quite different from those of larger immobilized cell supports. The microcarrier system used by Sinskey et al. (25) used beads of 160 μ in diameter and the total surface area was 24 cm^2/cm^3. Based on this information, gas-liquid mass and liquid-solid mass transfer rates were estimated to be 5.2 mmol/(L.h.atm) and 45 mmol/(L.h.atm). Since the gas-liquid mass transfer rate is much smaller than that of the liquid-solid, it is concluded that the gas-liquid mass transfer was rate controlling. The oxygen concentration at the microcarrier surface should be very close to that in the bulk liquid unlike the situation in example 1.

CONCLUSIONS

1) Oxygen transfer in immobilized whole cell systems can be analyzed with two or three step O_2 transport mechanisms and a rate-limiting step can be identified.

2) Penetration depths of oxygen in microbiol cell systems are estimated to be 50 - 200 μ and those of animal and plant tissue cultures are in the range of 500 - 1000 μ.

3) Oxygen transfer in a microcarrier system is usally gas-liquid mass transfer limited.

48

APPENDIX

Calculation of Oxygen Penetration Depth

For simplicity it is assumed that the oxygen consumption kinetic rate is zero-order and the cells are uniformly distributed throughout a spherical support shown below. Previously, oxygen penetration was studied with zero-order kinetics and other complex kinetics (32-34), but the effect of external mass transfer resistance was not taken into account.

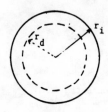

The oxygen mass balance and boundary conditions around the sphere are given by

$$D_{eff} \frac{1}{r^2} \frac{d}{dr} (r^2 \frac{d C_s}{dr}) = Q \qquad (A-1)$$

at $r = r_i$

$$k_f, eff (C_b^* - C_{bi}) = -D_{eff} \frac{d C_s}{dr} \bigg|_{r=r_i} \qquad (A-2)$$

$$C_s = C_b^* \qquad (A-2)'$$

at $r = r_d$

$$C_s = 0 \qquad (A-3)$$

$$\frac{d C_s}{dr} = 0 \qquad (A-4)$$

The application of equation (A-3) is necessary to obtain the penetration depth of oxygen beyond which oxygen is completey depleted. Solving equation (A-1) for C_s with equations (A-3) and (A-4) and substituting the result into equation (A-2), we obtain the following algebraic equation

$$k_f, \text{ eff } (C_b^* - \phi^2(\frac{1}{6} r_i^2 + \frac{1}{3} r_d^3/r_i - \frac{1}{2} r_d^2)) =$$

$$\frac{D_{eff}\phi^2}{3} (r_i - r_d^3/r_i^2) \qquad\qquad (A-5)$$

where $\phi = \dfrac{Q}{D_{eff}}$

r_d is obtained by trial and error from equation (A-5). If $r_d \leq 0$, oxygen penetrates all the way to the center of the particle. If $r_d > 0$, then the penetration depth is given by

$$P_d = r_i - r_d \qquad\qquad (A-6)$$

In the absence of external diffusion limitation equation (A-2)' is used instead of equation (A-2).

Subsequently total oxygen uptake rate (TOUR) is given by

$$\text{TOUR} = k_f, \text{ eff } (C_b^* - C_{bi}) a_s \phi_s \quad (m\mu \ O_2 /(L.h) \qquad (A-7)$$

where L refers to the unit volume of the reactor.

TOUR with infinite k_f is

$$(\text{TOUR})k_{f\to\infty} = -D \left. \frac{d \ C_s}{dr} \right|_{r=r_i} \qquad\qquad (A-8)$$

REFERENCES

1) D. Gbewonyo and D.I.C. Wang, Biotechnol. Bioeng., 25, 2873 (1983).

2) H.N. Chang, Y.-S. Kyung and B.H. Chung, Biotechnol. Bioeng. (in press).

3) H.N. Chang, I.S. Joo and Y.S. Ghim, Biotech. Lett., 6, 487 (1984).

4) S.F. Karel, S.B. Libicki and C.R. Robertson, Chem. Eng. Sci., 40, 1321 (1985).

5) J.D. Enzminger and J.A. Asenjo, Biotech. Lett., 8, 7 (1986).

6) R.A. Knazek, P.M. Gullino, P.O. Koller and R.L. Dedrick, Science, 178, 65 (1972).

50

7) D.S. Inloes, W.J. Smith, D.P. Taylor, S.N. Cohen, A.S. Michaels and C.R. Robertson, Biotechnol. Bioeng. 25, 2653 (1983).

8) C.R. Robertson and I.H. Kim, Biotechnol. Bioeng. 27, 1012 (1985).

9) C.R. Robertson and I.H. Kim, in Separation, Recovery and Purification in Biotechnology. Recent Advances and Mathematical Modelling. Asenjo, J.A. and Hong. H. (Eds.) (1986).

10) J. Briffaud and M. Engasser, Biotechnol. Bioeng., 21, 2093 (1979).

11) C.W. Lee and H.N. Chang, Enz. Microb. Tech., 7, 561 (1985).

12) W.D. Decker, in Fundamentals of Biochemical Engineering (H. Brauer, Ed.) p. 445, Biotechnology Vol. 2, Verlag Chemie, Heidelberg (1985).

13) H. Blenke, ibid, p. 465.

14) C. Webb, H. Fukuda and B. Atkinson, Biotechnol. Bioeng., 38, 41 (1986).

15) J.H. Kim, M. Hosobuchi, M. Kishimoto, T. Seki, T. Yoshida, H. Taguchi and D.D.Y. Ryu, Biotechnol. Bioeng., 27, 1445 (1985).

16) S.O. Enfors and B. Mattiasson, in Immobilized Cells and Organelles (B. Mattiasson, Ed.), Vol. II, p. 41 CRC Press, Boca Raton, Florida (1983).

17) J.M. Radovich, Enz. Microb. Tech., 7, 2 (1985).

18) M. Moo-Young and H.W. Blanch, in Adv. in Biochem. Eng. Vol. 19, p.1, A. Fiechter (Ed.), Springer-Verlag, Berlin (1981).

19) H.N. Chang and M. Moo-Young, submitted for publication (1986).

20) Y.T. Shah, B.G. Kelkar, S.P. Godbole and W.D. Deckwer, AIChE J., 28, 353 (1982).

21) W.D. Deckwer, in Adv. in Biotech., Vol. I, p. 471, M. Moo-Young, C.W. Robinson and C. Venzia (Eds.), Pergamon Press (1981).

22) R. Wittler, H. Baumgartl, D.W. Lubbers and K. Schugerl, Biotechnol. Bioeng., 28, 1024 (1986).

23) C.W. Lee and H.N. Chang, Biotechnol. Bioeng. (in press).

24) A. Margaritis, F.J.A. Merchant, CRC Crit. Rev. Biotech., 1 (4), 339 (1983).

25) A.J. Sinskey, R.J. Fleischaker, M.A. Tyo, D.J. Diard and D.I.C. Wang, Ann. N.Y. Acad. Sci., 369, 47 (1981).

26) D.I.C. Wang, "Fermentation Technology Lecture Notes", MIT Summer Course (1984).

27) J. Hopkinson, in Immobilized Cells and Organelles, vol. 1, p. 89, B. Mattiason (Ed.), CRC Press (1983).

28) K. Linsey and M.M. Yeoman, in Primary and Secondary metabolism of Plant Cell Cultures, Neumann et al. (Eds.), p. 304, Springer-Verlag, Berlin (1985).

29) F. Mavituna and J.M. Park, Biotech. Lett., $\underline{7}$, 637 (1985).

30) M.Y. Huang and H.R. Bungay, Biotechnol. Bioeng., $\underline{15}$, 1193 (1973).

31) H. Horitsu, S. Adachi, Y. Takahashi, K. Kawai and Y. Kawano, Appl. Microbiol. Biotech., $\underline{22}$, 8 (1985).

32) T. Yano, T. Kodama and K. Yamada, Agr. Biol. Chem., $\underline{25}$, 580 (1961).

33) T. Kobayashi, G. van Dedem and M. Moo-Young, Biotechnol. Bioeng., $\underline{15}$, 27 (1973).

34) J.A. Howell and B. Atkinson, Biotechnol. Bioeng., $\underline{18}$, 15 (1976).

A LARGE SCALE MEMBRANE REACTOR SYSTEM WITH DIFFERENT COMPARTMENTS FOR CELLS, MEDIUM AND PRODUCT

G. Klement,
W. Scheirer and H.W.D. Katinger

Institute of Applied Microbiology
Austria

INTRODUCTION

The cultivation of hybridomas and other animal and human cells on a large scale, still poses several problems today (TABLE 1). The

PROBLEM	SOLUTION	
Fragility to mechanical stress	Protection	Low power input and/or encapsulation
Genetic stability	Minimizing of total duplications Slowing down of growth speed	Saving of cell mass Physical or chemical limitation Space limitation
Low product concentration	Improvement of cell line Enrichment of product	Dislinking of the harvesting with the nutritive medium
Inactivation of product in the culture	Shortening of residence time	Increase cell concentration

protection of cells from mechanical stress is necessary because of their extreme fragility.

53

For this purpose, one can either minimize the power input into the reactor system or protect the cells by encapsulation in microcapsules or between membranes.

Some hybridoma lines show genetic instability which may be partially stabilized by slowing down of growth rate and by minimizing the total number of cell duplications. This can be achieved by retention of cell mass when harvesting the product, and by applying some kind of limitation, e.g., space limitation.

Cell lines which show low productivity may become useful by collection of the product over longer cultivation (and feeding) periods. Therefore, a dislinking of the product harvest with the nutrient medium is necessary. A degradation or inactivation of the product can be minimized by shortening the mean product residence time within the fermentor. This requires a maximum of cell concentration.

RESULTS

One possible solution will be a membrane system of three chambers, where the cells are immobilized between two different sheets of flat membranes (Fig. 1). The membrane between nutrient medium and the cells is an

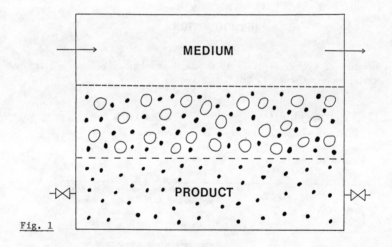

Fig. 1

ultrafilter with a molecular-weight cut-off which is useful for the particular product. It must be open enough to feed the cells with nutrients and provide them with oxygen by diffusion. The antibodies or other product must not pass this barrier and will remain within the cell compartment. The other side of this flat cell chamber is a microfiltration membrane of a pore size which is suitable to retain the cells from the product chamber, e.g. a 0.2 μ sterile filtration membrane. The product passes this membrane by diffusion into the product chamber. With this configuration, it is possible to feed the cells continuously and wait for a proper product concentration. This closely resembles the situation in mouse ascites culture. Similar cell densities $(80 \times 10^{6}/\text{ml})$ have been reached with mouse/mouse hybridomas. The production rate of antibodies is similar to static culture in flasks. The highest possible concentration of antibodies is dependent on the feed back regulation of the particular hybridoma line, and generally lies within the concentration range which can be reached in mouse ascites. The product harvest is performed repeatedly in proper time intervals by replacing the content of the product chamber with fresh medium without removing cells from the reactor. The purity of the product may be better than in mouse ascites because of the exclusion of large serum proteins by the feeding membrane barrier. For other processes, there is the possibility of using the third chamber for other purposes, e.g. oxygenation, induction, etc. To reach industrial scale with this system it is necessary to stack the units (Fig. 2). This is possible up to

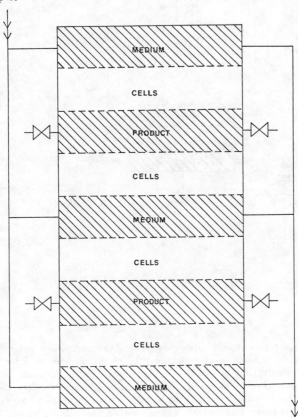

Fig. 2

several hundred layers such that each medium and each product chamber serves two cell chambers.

To achieve the proper environmental conditions for the cells, a periphery system is installed (Fig. 3). In principle it is the same as

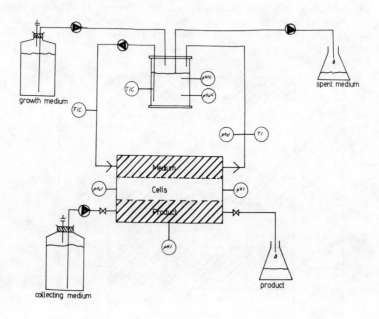

growth medium

spent medium

collecting medium

product

Fig. 3

with any other perfused immobilization systems. The conditioning vessel must be as small as possible for a short mean residence time of the medium in order to minimize inactivation and denaturation of nutrients. This is particularly important for media with low protein content. From this vessel, the medium is circulated through the system by a rotary pump; the circulation speed must be high enough to keep the oxygen tension at the outlet of the chamber high enough to avoid starving of cells. This can be measured by a probe which is mounted directly within the cell compartment. The perfusion rate of the medium compartment is dependent on the consumption of nutrients by the cells. It is controlled by an appropriate guide parameter,e.g. glucose concentration.

The commercial form of the system (Fig. 4) is sterilizable <u>in situ.</u>

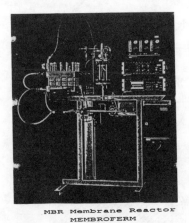

MBR Membrane Reactor
MEMBROFERM

Fig. 4

All connections can be resterilized during operation like any other cell culture fermentation plant.

The chambers are made of fluor carbon network and silicon gaskets (Fig. 5). The chambers are rinsable, have a height of 0.6 mm and a volumetric

Fig. 5

content of appr. 20 ml each. It is possible to use the whole system in a two - chamber operation mode, like a hollow fibre cartridge, but with a free choice of membranes and probably a better scaling-up potential.

Another operational mode is the use as a tube reactor (Fig. 6) which is

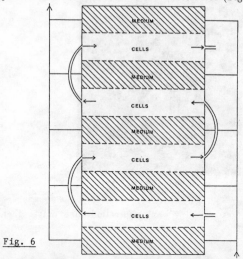

Fig. 6

possible by a slight modification of the gaskets. The new principle of this construction compared, to a conventional tube reactor, is that the tube is mantled by membranes, allowing refeeding of cells or enzyme treatment of material for bioconversion all the way through the 50 m long tube.

CONCLUSION

With this reactor system, it is possible to solve some of the problems in large scale cell culture technology. The system is of universal use and may be advantageous for the cultivation of cells which are fragile, unstable, have low production rates or give an unstable product. Some advantages can be seen in the use of the system for bioconversion. The real benefits can be estimated only after a broad evaluation, which remains to be done.

IMMOBILIZATION OF MULTIENZYME SYSTEM AND DEXTRAN-NAD$^+$ IN SEMIPERMEABLE MICROCAPSULES FOR USE IN A BIOREACTOR TO CONVERT UREA INTO L-GLUTAMIC ACID

Kang Fu Gu and Thomas Ming Swi Chang

Artificial Cells and Organs Research Centre
Faculty of Medicine, McGill University,
3655 Drummond Street, Montreal, P.Q., Canada

Most metabolic functions are carried out in biological cells by complex multienzyme systems. A major problem in using multienzyme systems is the requirement for cofactors [1]. Artificial cells based on semipermeable microcapsules have been used in the immobilization of enzymes, cells, microorganisms, cell organelles and other biologically active materials [2-6]. Cofactors covalently linked to soluble macromolecules can be microencapsulated. Because microencapsulated multienzyme systems and cofactors are all in a soluble form, there is no steric hindrance or diffusion restriction. Hence microencapsulation is an ideal immobilization method for multienzyme systems for basic research and biotechnological applications [2-5].

In our present study, semipermeable nylon-polyethylenimine microcapsules were prepared as described [6] based on the procedures of Chang et al. [2-9]. About 5 ml of microcapsules with a mean diameter of 80-100 μm were prepared in a batch preparation. The 5 ml of microcapsules contained 15 mg of L-glutamic dehydrogenase, 21 mg of yeast alcohol dehydrogenase, 1 mg of urease and 3 μmol of NAD$^+$ linked to 50-70 mg of dextran T-70. Dextran-NAD$^+$ was prepared according to the methods of Mosbach's group [10-12].

A substrate solution with the following concentrations was used: urea (20 mM), α-ketoglutarate (20 mM), ADP (0.1 mM), KCl (5 mM) and ethanol (200 mM).

Conversion of urea into L-glutamic acid by the L-glutamic dehydrogenase multienzyme system is shown schematically in Fig.1. L-glutamic dehydrogenase (GLDH), yeast alcohol dehydrogenase (YADH), urease and dextran-NAD$^+$ are retained within microcapsules. Urea and α-ketoglutarate diffusing into microcapsules are converted into L-glutamic acid. Ethanol acts as a cosubstrate for the regeneration of dextran-NADH.

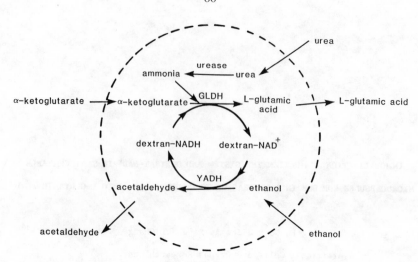

Fig. 1. Schematic representation of a microcapsule containing a multienzyme system and dextran-NAD⁺ for the continuous conversion of urea into L-glutamic acid.

For batch conversion, the microcapsule suspension to total reaction volume ratios ranged from 1:5 to 1:60. The microcapsules were kept in suspension during the reaction by shaking at 30 °C in a Lab-Line Orbit Environ-Shaker 18 at 130 rpm. From 22.6 to 53.4 μmol of L-glutamic acid were produced by 0.4 ml of microcapsules within 2 hrs. The corresponding conversion ratios were from 56.5% to 11.1%. If the reactions were continued for 6 more hrs, from 33.7 to 114.3 μmol of L-glutamic acid were produced. The conversion ratios increased from 84.3% to 23.8% (Table 1).

TABLE 1   CONVERSION OF UREA INTO L-GLUTAMIC ACID IN BATCH REACTION

Ratio	L-Glutamic acid (μmol)				Conversion ratio (%)			
	2 hrs	4 hrs	6 hrs	8 hrs	2 hrs	4 hrs	6 hrs	8 hrs
1:5	22.6	28.1	32.0	33.7	56.5	70.3	80.0	84.3
1:10	33.4	41.6	44.6	49.1	41.8	52.0	55.8	61.4
1:20	38.9	57.5	62.8	65.5	24.3	35.9	39.3	40.9
1:40	47.3	78.0	80.0	91.0	14.8	24.4	25.0	28.4
1:60	53.4	—	—	114.3	11.1	—	—	23.8

We have also used a continuous conversion bioreactor containing 4 ml of microcapsules. With a flow rate of SV=1.5 (hr^{-1}), the maximum conversion rate was 49.6 μmol of L-glutamic acid per hour. About 38% of the maximum conversion activity was still retained when continuously used for 4 days at 22 °C (Table 2).

61

TABLE 2   CONVERSION OF UREA INTO L—GLUTAMIC ACID IN A BIOREACTOR*

Fract.No.	Rate (μmol/hr)	R.A.** (%)	Fract.No.	Rate (μmol/hr)	R.A. (%)
1	2.42	4.9	30	26.01	52.4
2	16.94	34.1	40	25.40	51.2
3	49.60	100.0	50	24.80	50.0
4	44.76	90.2	60	19.96	40.2
5	43.55	87.8	70	19.96	40.2
6	43.55	87.8	80	20.57	41.5
7	42.34	85.4	90	19.96	40.2
10	32.66	65.9	98	20.57	41.5
20	30.85	62.2	107	18.75	37.8

* Flow rate was at about 6 ml/hr, with 27 fractions/day.
** Relative activity.

   Glucose dehydrogenase could effectively regenerate dextran—NADH within
microcapsules. Glucose dehydrogenase could be substituted for yeast
alcohol dehydrogenase in the L-glutamic dehydrogenase multienzyme system.
In this case, the conversion ratios were higher than that with yeast
alcohol dehydrogenase (Table 3).

TABLE 3   COMPARISON OF CONVERSION RATIOS*

Ethanol (200 mM)		Glucose (100 mM)		Glucose (200 mM)	
2 hrs	8 hrs	2 hrs	8 hrs	2 hrs	8 hrs
56.5+0.0%	84.3+1.7%	52.5+4.3%	96.2+1.0%	62.4+1.7%	96.2+1.0%

*The ratio of microcapsule suspension to reaction volume was 1:5.

   In summary, urea could be converted into L-glutamic acid by
microcapsules containing an L-glutamic dehydrogenase multienzyme system.
Yeast alcohol dehydrogenase and glucose dehydrogenase could effectively
regenerate dextran—NADH within microcapsules. L-glutamic acid formed from
urea could be further converted into L-alanine by the addition of
transaminase to the L-glutamic dehydrogenase multienzyme system in the same
microcapsules [7,13]. Urea or ammonia can also be converted into other
different amino acids with microcapsules. This approach may bring good
prospects for biomedical and industrial applications in the future.

ACKNOWLEDGMENTS

The support of the NSERC strategic grant and the MESST Virage Centre of excellent grant in biotechnology to TMSC is gratefully acknowledged. The graduate student fellowship of KFG from the MESST Virage Centre Grant is also acknowledged. The technical assistance of Mr. Colin Lister is much appreciated.

REFERENCES

1. Gestrelius,S., Mansson,M.O. and Mosbach,K. (1975), Eur. J. Biochem., 57, 529.
2. Chang,T.M.S. (1964), Science, 146, 524.
3. Chang,T.M.S. (1972), Artificial Cells, Charles C Thomas Co., Springfield, II.
4. Chang,T.M.S. (1984), Appl. Biochem. Biotechnol., 10, 5.
5. Chang,T.M.S. (1977), Biomedical Applications of Immobilized Enzymes and Proteins, Vol.1&2, Plenum Press, New York, N.Y..
6. Chang,T.M.S. (1985), Methods in Enzymology, Vol.112, (Widder,K.J. and Green,R. ed.), Academic Press, INC., New York, p.195.
7. Grunwald,J. and Chang,T.M.S. (1979), J. Appl. Biochem., 1, 104.
8. Grunwald,J. and Chang,T.M.S. (1981), J. Mol. Catal., 11, 83.
9. Yu,Y.T. and Chang,T.M.S. (1982), J. Enzyme Microb. Technol., 4, 327.
10. Mosbach,K., Larsson,P.O. and Lowe,C. (1976), Methods in Enzymology, Vol.44, (Mosbach,K. ed.), Academic Press, New York, p.859.
11. Lindberg,M., Larsson,P.O. and Mosbach,K. (1973), Eur. J. Biochem., 40, 187.
12. Larsson,P.O. and Mosbach,K. (1974), FEBS Lett., 46, 119.
13. Chang,T.M.S., Molouf,C. and Resurreccion,E. (1979), Artif. Organs, 3, 284.

Development of a high capacity adsorbent for enzyme isolation and immobilization

C. Dauner-Schütze[1], E. Brauer[1], A. Borchert[2], K. Buchholz[3]

1) Universität Frankfurt, Institut für Physikalische und Theoretische Chemie
2) Fachhochschule Ostfriesland, Emden
3) Institut für landwirtschaftliche Technologie und Zuckerindustrie an der TU Braunschweig

Biocatalyst preparation and production from microbial enzymes requires several steps or unit operations respectively:

> Fermentation
> Separation
> Concentration, purification
> Immobilization

We have developed an adsorbent with high capacity for technical application in enzyme concentration and purification.

We furthermore present a method for integrating the before mentioned steps of concentration, partial purification and immobilization in a combined procedure.

It is well known that bentonite and other clay minerals exhibit excellent adsorption capacities (1)(2). They have been applied in various processes on an industrial scale (see e.g. 3). It has, however, serious disadvantages residing mainly in difficult handling due to its very low and inhomogeneous particle size (5 to 30 μm). Therefore handling in industrial separation processes is difficult.

Our aim was to prepare particles on the basis of bentonite with appropriate diameter and shape and to maintain essentially the high capacity of the original material.

1 Entrapment of bentonite in polymer matrices

Adsorption isothermes exhibit the main characteristics of original bentonite (measurements with penicillin acylase (PA); similar results have been obtained with trypsin):

- high adsorption capacity 12000 U(PA)/g(bentonite),
  1,5 g(protein)/g(bentonite)

- unusual adsorption isotherm with maximum (c.f. 1)

- preferential adsorption of active enzyme from protein solution

Electrostatic interaction plays a key role in adsorption. Thus desorption of enzymes can be performed simply by shift of pH and ionic strength essentially without loss in activity. The adsorbent, however, cannot be reused in general.

The unusual isotherm possibly is due to the flexibility of the bentonite lattice (layers) (1), protein-protein and proten-bentonite interaction.

<u>Figure 1:</u> a) Adsorption isotherm for penicillin acylase
(x : Na-bentonite, o: Ca-Bentonite; pH 5, T 25°C)
b) Adsorption isotherm for total protein

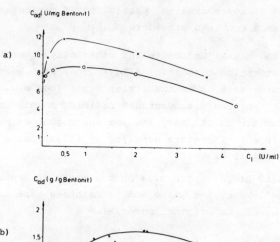

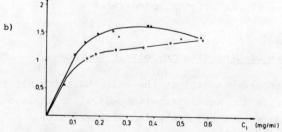

In order to obtain bentonite particles appropriate for reuse
in batch stirred tank processes or columns, e.g. in fluidized
bed operation, the original material must be entrapped in polymer matrices, similar to entrapment of microorganisms (c.f. 4, 5).
For this purpose bentonite is first equilibrated in buffer appropriate for protein adsorption (e.g. pH 5, 0,3 g/l NaCl or
0,55 g/l $CaCl_2$). A suspension of bentonite (10 to 100 mg/ml)
is then mixed with a solution of carrageenan or chitosan
(1 - 8 %). After intense mixing the suspension is dropped by
means of capillaries into a solution with appropriate counterions
(e.g. $CaCl_2$, 20 g/l for carrageenan or Na-polyphosphate, 15 g/l,
pH 8 for chitosan). Ionotropic gelation gives nearly spherical
particles, with the particle diameter depending on the internal
diameter of the capillaries and the pressure applied (c.f. 4 - 7)
(Fig. 2).

Figure 2: Particles with bentonite obtained after ionotropic
gelation with carrageenan

2 Adsorption characteristics of entrapped bentonite

The particles in figure 2 have been exposed to shear stress in a laboratory stirred vessel for 16 h where they exhibit sufficient mechanical stability.

Adsorption capacities after entrapment are lower than with original bentonite, yet very high absolute values have been obtained as given in table 1.

Table 1: Adsorption capacities of entrapped bentonite preparation for crude penicillin acylase (gift from Hoechst AG, Frankfurt/Main)

Preparation	Composition	Adsorption Capacity (U/g(prep. dry weight))	(U/g(bentonite)
bentonite/chitosan	2 : 1	1195	1780
bentonite/chitosan	5 : 1	2025	2445
bentonite/carragee-nan	1 : 1	7000	3000
bentonite/carragee-nan	5 : 1	6500	5500
carrageenan	–	12000	–
bentonite	–	–	12000

The capacity for protein adsorption is also high, e.g. 0,9 g (protein) per g carrier for bentonite/carrageenan 5 : 1. At low equilibrium concentrations penicillin acylase is preferentially adsorbed.

Desorption can be achieved by increasing pH (shift from pH 5 to pH 8) and ionic strength. Preferential desorption of enzyme can be favoured by gradual desorption where the enzyme is desorbed first. Kinetics of adsorption are diffusion limited.

Adsorption isotherms for entrapped bentonite particles are shown in figure 3. Three results are obvious from these investigations:

- the high adsorption capacity which is also due to adsorption on the carrageenan matrix,

- the preferential adsorption of active enzyme (PA) at low
  equilibrium concentration,

- normal form of the isotherms (without a maximum),

Figure 3: Adsorption isotherms for PA (a) and total protein
(b) on bentonite/carrageenan $0 : 1$ (x), $0,2 : 1$ (Δ),
$1 : 1$ ($\square$), $5 : 1$ (o)

a)

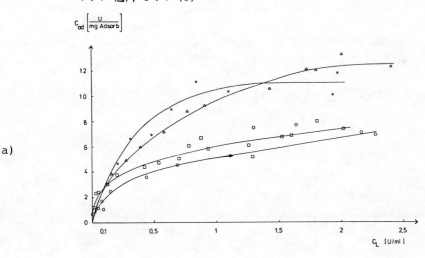

b)

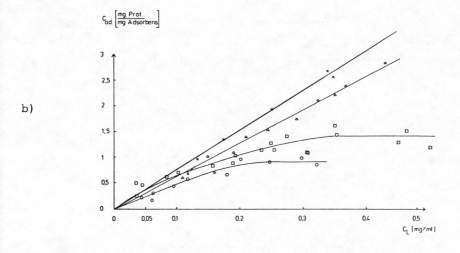

68

Logarithmic plots obey a Freundlich-type-isotherm (Fig. 4).
They are in distinct contrast to those of original bentonite
with unusual isotherms where a maximum for adsorbed penicillin
acylase in an intermediate range of equilibrium concentration
has been observed (Fig. 1). From this it can be concluded that
the flexibility of the crystalline bentonite layers which has
been studied extensively with aminoalkans (1) is an important
factor for the interpretation of the unusual adsorption phenomena observed with bentonite. After entrapment the flexibility of the bentonite layer should be essentially restricted.

Figure 4: Logarithmic plot of adsorption isotherms for PA,
bentonite/carrageenan 1 : 1 (x) and 5 : 1 (o)

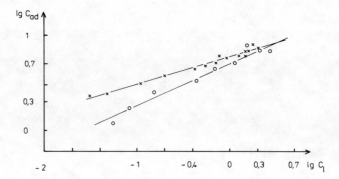

69

3 Combined enzyme isolation and immobilization

Due to the favourable adsorption characteristics enzymes
(e.g. PA) can be adsorbed preferentially from dilute crude
protein solution with high enzyme loading on bentonite. This
can be processed straight forward to obtain an immobilized biocatalyst by crosslinking the enzyme with glutardialdehyde (0,5 %
in solution for about 15 min). The complex can then be entrapped
in a polymer matrix as described before in order to obtain appropriate biocatalyst particles.

Good results have been obtained with alginate as polymer
(4 % solution) which can be precipitated after mixing with
bentonite/penicillin acylase in $CaCl_2$ solution (1 %). The
residual activity of PA was 66 %, no attrition was observed
in laboratory scale tests in stirred vessels (in 50 mmol/l
$CaCl_2$). Entrapment in chitosan gave biocatalysts which were
also stable in phosphate buffer with an activity of 520 U/g
(dry biocatalyst). No loss in activity was observed during
400 h of operation (penicillin hydrolysis).

Further optimization of the biocatalysts seems possible, e.g.
concerning effectiveness by means of optimal carrier design
(c.f. 8, 9).

Utilization of entrapped bentonite with appropriate particle
dimension thus offers several option in enzyme isolation and
biocatalyst design:

- high adsorption capacity for enzyme adsorption from dilute
  solutions,

- convenient separation and reutilization,

- application in fluidized bed reactors (c.f. 10),

- combination of enzyme isolation and immobilization.

These investigations have been funded by a grant from BMFT (Federal ministry for research and technology).

References

1   Rodriguez, J.L.P.; Weiss, A.; Lagaly, G.: Clays Clay Miner. 25
    (1977) 243 - 251

2   Armstrong, D.E.; Chesters, G.: J. Soil Sci. 98 (1964) 39

3   Kutzbach, C.: DOS 2217 745 (1972)

4   Vorlop, K.D.; Klein, J.: Biotechnol. Lett. 3 (1981) Nr. 1,
    9 - 14

5   Klein, J.; Wagner, F.; Vorlop, K.D.: Dechema-Monogr. 84 (1979)
    265 - 273

6   Dauner, C.; Borchert, A.; Buchholz, K.: Patent application
    P 3440 4449-43 (1984)

7   Dauner, C.; Borchert, A.; Buchholz, K.: Chem.-Ing.-Techn. 58
    (1986) 491 - 493

8   Borchert, A.; Buchholz, K.: Biotechnol. Bioeng. 26 (1984)
    727 - 736

9   Buchholz, K.: Adv. Biochem. Eng. Vol. 24 (1982) 39 - 71

10 Hicketier, M.; Borchert, A.; Buchholz, K.: Chem.-Ing.-Techn. 57
    (1985) Nr. 5, S. 449

NYLON FILTERS WITH RENNET ENZYME (CHYMOSIN) FOR CONTINUOUS MILK-CLOTTING

L. AMOURACHE and M. A. VIJAYALAKSHMI*

Dept. of Bioengineering, Univ. Technol. de Compiegne,
France

*Dept. Food Science, Universite Laval, Quebec, Canada

INTRODUCTION

The scarcity and the high cost of milk clotting enzymes, traditionally obtained from the kid or calf abomasa and used in batch systems, have encouraged the research for alternatives. One of them is the search for microbial substitutes and the second being an immobilized enzyme system (1-4). The first one has the disadvantage of unacceptable organoleptic properties. Hence, the second alternative looks more attractive.

Though an immobilized enzyme is successfully used in the case of invertase for high fructose syrup, the successful utilisation of immobilized enzyme in the case of macro molecular substrates is still rather a challenge.

The reasons are:

- mass transfer limitations,

- complexity of the macro molecular substrates like pectin milk proteins, etc.

In both the examples cited, the mechanism is an hydrolytic one which is enzyme triggered followed by electrostatic non-enzymatic mechanism. Nylon filters could be good alternatives for the chymosin immobilization, especially for a two step continuous milk clotting procedure.

71

EXPERIMENTAL

The high resistance nylon filters of 3 mesh size 1250μm.; 2800 and 4000 are from UGB, France. Glutaraldehyde, carbodiimide (EDC) as well as all other reagents are from MERCK. The enzyme used is a kid abamasom extract as described by (6).

Preparation of the nylon support:

The nylon filter used is a polyamide 6 whose amide bond is partially hydrolysed by 3 M HCl during 4 hours according to (5), in order to liberate enough free NH_2 and COOH reactive groups, without destroying the physical structure.

Activation of the nylon filters:

Before coupling the enzyme, the partially hydrolysed nylon filters are activated either by glutaraldehyde or by carbodiimide (CDI) according to (7).

Coupling of the enzyme:

The activated nylon filters are washed with 0.001 M Sodium acetate buffer at pH 5.5.

To 0.13 g (equivalent to 3 cm^2) 2 ml of an enzyme solution are added at $+4\degree$C. and kept stirring for 24 hours. Then the enzyme coupled nylon filters are washed with 0.01M Sodium acetate buffer pH 5.5, before using.

The enzyme solution used has the following characteristics:

Specific activity = 1222 SU/mg protein
Protein content = 9 mg/ml

Enzyme activity assay:

A 12.% reconstituted solution of the spray dried milk in 0.01 M $CaCl_2$ is used as substrate. The coagulation is performed at $+37\degree$C. and the time of the appearance of the clot on the sides is an indication of the activity. Times between 1 and 3 min. for the appearance of the clot are taken as valid. The quantity of nylon filters containing the enzyme are adjusted to be in this range.

The activity is calculated in soxhlet units as follows:

(SU) = 2400/T x 5/0.5 x D
where (SU) = soxhlet units; T = time for coagulation and D = dilution factor.

A Semi continuous reactor:

The semi continuous milk clotting reactor system is achieved by inserting 5 discs of 1 cm^2 diameter and corresponding to a weight of 400 mg into a glass column. Each disc is separated from the other by a glass ring. A 12% milk powder solution in 0.01 M

$CaCl_2$ is pumped through the column. The milk reservoir as well as the column are kept at $+4°$ C. by circulating cold water in a jacket. Parallel tubes containing 4.5 ml of the untreated milk are incubated in a water bath at $+37°$ C. Then 0.5 ml of the enzyme treated milk, from the outlet of the column are pumped into each tube and the clotting time for each tube and hence the activity in soxhlet units is calculated.

RESULTS

<u>Choice of the nylon filters:</u>

Table 1 summarises the coupling yield of the active enzyme on the nylon filters of different mesh size and according to three different methods of coupling the enzyme, simple adsorption, coupling through glutaraldehyde and through CDI. It could be seen that the nylon 2800 and 4000 give reasonable coupling yields, even with the simple adsorption technique, (133 SU/cm^2 of nylon). The nylon 1250 gives very low yields. However the non-covalent adsorption, as expected, results in rather poor stability of the immobilized enzyme.

Coupling Method	Nylon Type used	% of Protein coupled	% Immobilized Activity/Protein coupled	Residual activity after 5 cycles as % initial activity
Adsorption	1250	–	negligeable	–
	2800	–	10.0	16
	4000	–	10.0	40
with glutaraldehyde	1250	–	1.5	–
	2800	44.4	11.1	17
	4000	30.0	11.6	40
with carbodiimide	1250	–	–	–
	2800	–Nd–	12.5	75
	4000	–Nd–	16.6	28

Nd: not determined

Table 1: CHYMOSIN COUPLING YIELD WITH DIFFERENT TYPES OF NYLON FILTERS AND WITH DIFFERENT COUPLING METHODS

In the case of covalent coupling of the chymosin, the carbodiimide condensation gives the best yield (as much as 17.0%), with good stability. In particular the enzyme coupled to nylon 2800 via carbodiimide conserves as much as 75% of its initial activity after 5 cycles of milk clotting assays. Based on these, further results will be focussed on the 2800 mesh nylon only.

Optimisation of the covalent coupling of the chymosin

Coupling via glutaraldehyde:

Glutaraldehyde is a bifunctional reagent and reacts with NH_2 groups. The coupling of a protein to nylon with the aid of glutaraldehyde involves the free NH_2 of the polyamide liberated by the controlled acid hydrolysis. Hence, it is important to use an optimal concentration of this bifunctional reagent to get a good coupling yield, without inactivating the enzyme by inter, intra molecular bridging. Table 2 shows the activities tested on nylon 2800 as a function of glutaraldehyde concentration used. 0.06% glutaraldehyde in a mixture containing 9 mg of enzyme and 0.13 g of nylon gives the best results.

Glutaraldehyde conc. (%)	0	0.02	0.04	0.06	0.08	0.1
Total Activity coupled (SU)	1100	857	888	1100	858	0

Table 2: EFFECT OF GLUTARALDEHYDE CONCENTRATION ON THE ENZYME ACTIVITY.

Table 3 shows the effect of enzyme concentration in the coupling medium on the coupling yield. There is a direct relation between the coupling yield and the enzyme concentration up to 9 mg of enzyme in the mixture and the increasing concentrations above this level are without any effect.

	Nylon 2800				
Protein conc. in the reaction mixture	1.5	7	9	11	13
Protein coupled on nylon	0.5	3	4	4	4
Active enzyme coupled to nylon SU/cm^2	114	113	177	177	177

Table 3: EFFECT OF ENZYME PROTEIN CONCENTRATION IN THE REACTION MIXTURE ON THE ACTIVE ENZYME–NYLON COMPLEX YIELDS.

Coupling through Carbodiimide condensation

The reaction of proteins with glutaraldehyde is rather an ill-defined process and the actual mechanism is not known. The peptide condensation with the aid of carbodiimide between the free COOH groups of the nylon and the free NH_2 of the protein could be a good alternative to obtain a stable immobilized chymosin preparation.

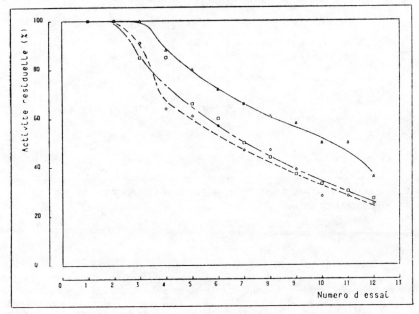

Figure 1:   Stability of the immobilised enzyme as a fonction
             of carbodiimide concentration used for coupling.

             (◦- - -◦) carbodiimide 0,08 M

             (◦——◦) carbodiimide 0,1 M

             (◻— -◻) carbodiimide 0,2 M

Figure 1 and Table 4 show the effect of varying carbodiimide concentrations, both on the yield of the coupling and its stability with repeated utilisation. It could be seen that 0.1 M carbodiimide gives the best yield and also the maximum residual activity after twelve successive milk clotting activity assays.

76

MOLAIRE CONCENTRATION CARBODIIMIDE	NYLON 2800 HR					
	0,008		0,1		0,2	
No. ESSAI	ACTIVITY SU/cm^2	ACTIVITY RESIDUEL (%)	ACTIVITY SU/cm^2	ACTIVITY RESIDUEL (%)	ACTIVITY SU/cm^2	ACTIVITY RESIDUEL (%)
1	145	100	200	100	133	100
2	145	100	200	100	133	100
3	133	91	200	100	114	85
4	94	64	177	88	114	85
5	88	61	160	80	88	66
6	84	57	145	72	80	60
7	69	47	133	66	66	57
8	69	47	123	61	59	44
9	57	39	114	58	50	37
10	41	28	88	50	44	33
11	41	28	88	50	40	30
12	36	25	72	36	36	27

Table 4: COUPLING YIELD AND STABILITY AS A FONCTION OF CARBODIIMIDE CONCENTRATION USED.

Stability of the different preparations

Table 1 shows the comparative stabilities of the different preparations of the immobilized chymosin prepared according to the three methods namely simple absorption and covalent coupling either with glutaraldehyde or with carbodiimide. The carbodiimide coupling gives a relatively more stable preparation. Only 20 to 30% of the initial activity is lost after five successive utilisation, whereas almost 80% of the initial activity is lost in the case of glutaraldehyde coupling.

Paquot et al., (1976) have already reported that addition of Bovine Serum Albumin (BSA) to the immobilized enzyme complex could improve the stability. In our hands, the BSA had a positive effect only in the case of coupling through glutaraldehyde (Fig. 2) and it is without effect on the preparation obtained by carbodiimide condensation. This implies that the free carbonyl groups on the nylon could be at least partially responsible for the loss of activity. It is well known that milk clotting with chymosin involves two steps; one enzymatic and the second electrostatic. Hence, it is highly plausible that the charged groups on the surface of the matrix can influence the activity. Also it has been suggested by Ferrier et al., (8) that even at low temperatures some coagulation does occur and the resulting whey proteins, $CaPO_4$ and some factors in the micellar materials are chymosin inhibitors. The adhesion of these materials to the nylon can provoke the observed loss in enzyme activity.

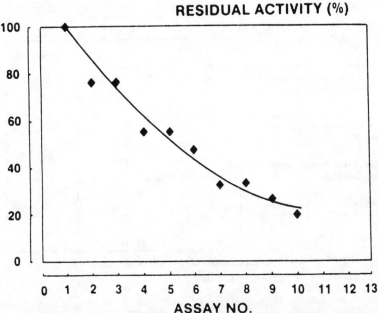

RESIDUAL ACTIVITY (%)

ASSAY NO.

Fig. 2 Protection by a 1mg/ml BSA solution of the immobilised chymosin activity.
Note: The residual activity is enhanced from ~ 18.1 to 50% at the 5th successive assay.

In the case of coupling chymosin to nylon with glutaraldehyde, the final product will have more residual free COOH groups on the surface of the matrix and the BSA can in a way neutralize these charges, which otherwise could contribute to the inactivation of the enzyme. It is also possible that the added BSA neutralizes the excess glutaraldehyde.

Apart from the loss of activity observed with successive utilisation of the immobilized chymosin, we studied their shelf life. The shelf life was found to be influenced by the salt concentration added to the solution in which the preparation was stocked. Table 5 shows the effect of a 24 hours storage of the complex in solution of different NACl concentrations.

From these data, it is very clear that there is a salting out effect at 10 and 100 g NaCl/L, by which almost the totality of the coupled enzyme is found leached out into the solution. Concentration of 500 and 1000 g NaCl/L seem to stabilize the enzyme during storage, as no activity could be detected in the solution. However, the contact with the substrate does desorb up to 10% of the coupled enzyme, as this activity could be detected in the whey after the milk clotting. We observed good stability with NaCl concentration of 500 g/L and 1000/L, as stocking solutions. It was interesting to follow the stability over a longer period of time.

	NaCl concentration g/litre			
	10	100	500	1000
Activity at the (day 1) start (SU/cm^2)	133	133	126	133
Activity still bound to nylon after 24 hrs storage (SU/cm^2)	2.5	8	95	106
Activity in the stock- ing solution SU	124	100	0	0
Activity in the whey after milk clotting SU	0	0	10	8

Table 5: EFFECT OF NaCl CONCENTRATION IN THE STOCKING SOLUTION, ON THE STABILITY OF THE BOUND ENZYME.

Figure 3a shows evolution of residual activity with time of the chymosin-nylon complex (coupled by carbodiimide condensation) stored in 500 g/L and 1000 g/L NaCl solution after each enzyme activity assay. A 500 g/L NaCl solution seems to be the best as stocking solution. When we compare the residual activity observed with a preparation stored in a solution without NaCl, we can see that after 12 days, 40% of the initial activity is preserved whereas only 25% in the latter case.

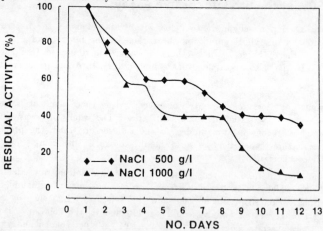

Fig. 3a   Effect of NaCl on the shelf lie of the immobilized enzyme.

A semi continuous reactor for milk clotting

In spite of the relative stability we can observe in Figure 3b that the enzyme-nylon complex is more stable after the initial phase of inactivation. Based on this we devised a reactor which is schematically represented in Figure 4. This functions on a two step milk clotting device similar to the one proposed by Cheryan et al (1). The nylon-enzyme complex is used for milk clotting for 12 to 15 successive cycles before inserting it into the reactor, so that only the stabilized form is used as indicted by Cheryan et al (1).

The influence of flow rates and the pH of the substrate are studied. Figure 5 shows the effect of flow rate on the milk clotting activity determined in two steps after dilution as described in the experimental section. Little difference in milk clotting activity is observed for flows above 7 ml/min. This implies that only a partial enzymatic hydrolysis in the first step is necessary to trigger the coagulation in the second step. This is in agreement with Cheryan et al (1).

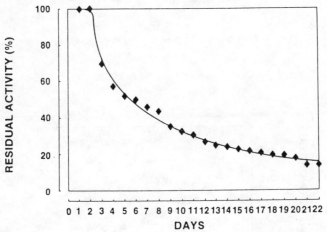

Fig. 3b Stability as a function of time (days) of the chymosin immobilized through carbodiimide.

At present we do not know precisely what is the minimum degree of hydrolysis necessary for the secondary coagulation phase. We are trying to determine this with the aid of an HPLC-gel filtration method (9).

Figure 6 shows the effect of the pH of the substrate used for the two step coagulation, on the efficiency of the enzyme reactor. At pH 5.5 we find the activity to be constant, whereas at higher pH's the loss of activity increases with even a slight increase in the pH. A rise from pH 6.2 to pH 6.8 makes the enzyme totally inactive, within 15 min. Ferrier et al., (8) have already observed a similar pH effect in immobilized pepsin used for the milk clotting. Green, M. L. and Crutchfield, G. (1969) (10) have also shown that, in spite of the fact that the rennet enzyme is stable up to pH 7.5 the milk clotting activity decreases almost exponentially above pH 6.2 even with soluble enzymes.

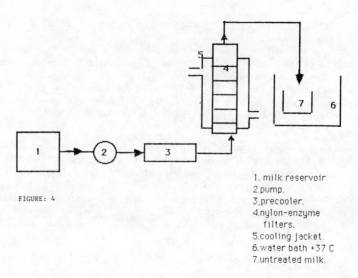

FIGURE: 4

1. milk reservoir
2. pump.
3. precooler.
4. nylon-enzyme
   filters.
5. cooling jacket.
6. water bath +37 C
7. untreated milk.

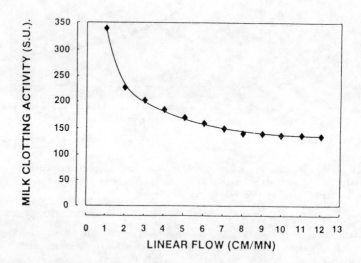

Fig. 5 Flow rate dependence of the milk clotting

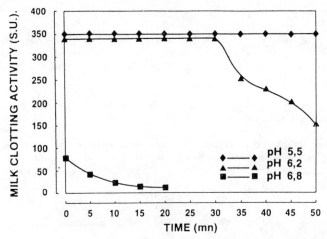

Fig. 6 Milk clotting activity of the reactor system as a fonction of substrate pH.

DISCUSSION

The immobilization of chymosin is a real challenge. Whatever the method used we find an instability of the immobilized enzyme. This has been observed by all the previous workers (1-4) Voutsinas and Nakai (4) argue that hydrophobic adsorption could be more efficient, but even then the stability was not very good. From our studies we see that the immobilization of chymosin, to partially hydrolysed nylon, through carbodiimide condensation gives a comparatively stable preparation. Moreover, the open mesh nylon filters and the reactor with well spaced filters as described could minimize fouling, which is one of the causes for the loss of enzyme activity even at low temperatures (Ferrier et al., (8). Also, using the enzyme activity at 4°C. instead of the 15°C. as described by Ferrier et al., 1972 (8) and Cheryan et al., 1979 (1) further minimizes the fouling. In spite of the low activity at this temperature, we can see that it was enough to trigger the coagulation at 37°C. in the secondary phase.

The persistant instability denotes that the immobilization was not totally covalent. A reasonable amount of the immobilized chymosin is simply adsorbed to the partially hydrolysed polyamide and held by electrostatic as well as hydrophobic forces, as the NaCl seems to stabilise in this case.

To overcome the residual charge effects, a four component condensation which neutralizes both COOH and NH_2 of the nylon using UGI reaction to immobilize the enzyme, as in the case of pectinesterase (Vijayalakshmi et al., 5) could be a good alternative.

The leaching could not be the only cause of the loss of activity. A partial oxidation and hence conformational changes of the bound enzyme can result in denaturation (suggestion from R. Messing). However, we could not observe any significant stabilisation by the addition of reducing agents such as DTT.

82

CONCLUSION

1. Though the immobilised chymosin system has limited stability this can be successfully used with chymosin substitutes where the limited enzymatic attack of the casein is needed to avoid the formation of bitter peptides resulting from prolonged hydrolysis.

2. The low working temperature for the first catalytic phase could minimise any microbial contamination.

3. Apart from technological application, this system could be useful in localising and studying the relation between K-casin degradation and the coagulation formation due to the separation of the enzyme phase from the milk clotting phase.

REFERENCES

1. M. Cheryan, P. J. Van Wyk, N. F. Olson and T. Richardson, J. Dairy Sci. 58(4): 477-481. 1975.

2. M. Paquot, Ph. Thonart and C. Deroanne. Le lait, 553; pp. 154-163. 1976.

3. M. J. Taylor, M. Cheryan, T. Richardson and N. F. Olson, Biotech and Bioeng. 19: 683-700. 1977.

4. L. P. Voutsinas and S. Nakai. J. Dairy Sci. 66: 694-703. 1982.

5. M. A. Vijayalakshmi; D. Picque; R. Jaumouille and E. Segard. In: "Food Process Eng." Vol. 2 pp. 152-158. 1979. Ed. P. Linko and J. Larinkari.

6. L. Amourache et M. A. Vijayalakshmi, J. Chromatogr. 303. (1984) 285.

7. P. V. Sundaram and W. E. Hornby. Febs lett; 10. (1970) 320.

8. L. K. Ferrier; J. Richardson; N. F. Olson and C. L. Hicks. J. Dairy Sci. 55. (6); 726-734, 1972.

9. M. A. Vijayalakshmi; L. Lemieux and J. Amiot. J. of Liquid Chromatography, 9. (16): 3559-3576, 1986.

10. M. L. Green and G. Crutchfield. Biochem. J. 115. 183, 1969.

11. R. Messing (personal communication).

The Development of an Immobilized Heparinase Reactor

by
Victor C. Yang[a], Howard Bernstein[b],
Charles L. Cooney[b], and Robert Langer[c] [d]

[a]College of Pharmacy, University of Michigan, Ann Arbor, Michigan 48109

[b]Department of Chemical Engineering, Massachusetts Institute of Technology, Cambridge, Massachusetts 02139

[c]Department of Applied Biological Sciences, Massachusetts Institute of Technology, Cambridge, Massachusetts 02139

[d]Department of Surgery, Children's Hospital Medical Center, Boston, Massachusetts 02115

ABSTRACT

Extracorporeal blood circulation has been used in many clinical situations such as kidney dialysis, cardiac surgery, plasmapheresis, and organ transplantation. It requires heparin anticoagulation to provide blood compatibility. However, systemic use of heparin results in a high incidence of bleeding complications. To solve this problem, we suggest a novel approach which would permit full heparinization of blood entering the extracorporeal devices but would enable enzymatic elimination of the heparin before the blood is returned to the patient. This approach consists of placing a reactor containing immobilized heparinase, a heparin-degrading enzyme, at a position after the extracorporeal device.

This paper will discuss the development of such a heparinase reactor. It will include the methods of heparinase production, purification, immobilization, reactor design, and preliminary data regarding the in vivo testing of the heparinase reactor.

83

Extracorporeal blood circulation(ECBC) has been employed in many clinical situations such as renal dialysis, open heart surgery, blood oxygenation, plasmapheresis, and organ transplantation. It requires anticoagulation to prevent the clotting processes that are initiated when blood comes into contact with extracorporeal devices (Figure 1A). Unfortunately, systemic heparinization leads to a high incidence of hemorrhagic complications (1-3).

A. Current Extracorporeal Circulation

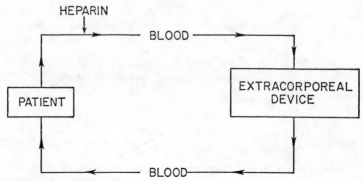

Figure 1A. Schematic diagram of the current extracorporeal circulation.

Such complications are more severe in acutely ill patients already at high risk of bleeding (e.g. patients who have undergone surgery or who have experienced severe trauma) (4,5),and in patients on the extracorporeal devices for extended time,e.g. patients on the blood oxygenator) (6). A number of approaches have been explored to solve this life-threatening problem associated with the systemic use of heparin. These approaches include the development of heparin substitutes (e.g. prostacyclin, low molecular weight heparins) (7-9); the development of blood compatible materials (e.g. materials with heparin or other substances bound to their surface) (10-11); and the administration of anit-heparin drugs such as proamine (12). In spite of these efforts, systemic heparinization continues to be used in all ECBC procedures, and control of blood heparin levels remains a serious problem to human safety.

We have suggested a novel approach which would permit full heparinization in the extracorporeal device, and yet would enable elimination on demand of heparin in the patient's bloodstream. This approach consists of placing a blood filter containing immobilized heparinase, an enzyme which degrades heparin into nearly inactive low molecular weight fragments, at the effluent extracorporeal device (Figure 1B). Such a filter could theoretically be used to eliminate heparin after it served its purpose in the extracorporeal device and before it is returned to the patient.

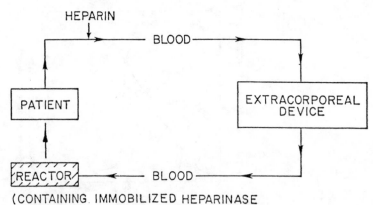

Figure 1B. Schematic diagram of the proposed extracorporeal circulation.

In this paper we review our efforts regarding the development of the heparinase reactor. The areas to be discussed herein include assays for heparin and heparinase, heparinase production, heparinase purification and characterization, heparinase immobilization, reactor design, in vivo testing of the reactor, and toxicology studies of the reactor and the heparin degradation products.

ASSAYS

Protein concentrations were measured by the Lowry method (13).
Heparinase activity was measured in vitro in two ways: (i) the disappearance of heparin, and (ii) the appearance of heparin degradation products. The former is measured by the metachromatic shift of Azure A from blue to red in the presence of heparin (14-15), while the latter is measured by the increase in ultraviolet absorption at 232 nm due to the formation of unsaturated uronides (14-15). One unit of heparinase activity is defined by the amount of enzyme which degrades 1 mg of heparin per hour.

For the in vivo blood deheparinization studies, the loss of heparin's anticoagulant activity is followed by both the APTT clotting assay (15-16), and anti-FXa amidolytic assay (15-16).

HEPARINASE PRODUCTION

Heparinase was produced from a gram negative rod shaped (1.0 x 0.3 um), non-motile, and non-spore forming soil bacterium, Flavobacterium heparinum. In previous methods of heparinase production, the bacteria was grown in an undefined complex protein digest medium, with heparin employed as the inducer (17). The fermentation was quite expensive because of the cost of heparin, and it also resulted in a low and irreproducible heparinase production (17,18). Heparinase was produced slowly at a level of 9,6000 units per liter of fermentation broth, and the volumetric productivity was usually maintained at 375 units per litre of broth per hour (Figure 2A) (18). In

addition, a rapid

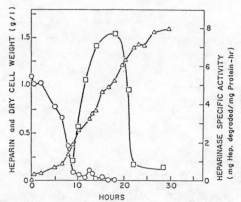

Figure 2A. Results of a typical fermentation in undefined complex protein digest medium showing heparin (O),
heparinase specific activity (□) and dry cell weight (△)
as a function of time in a 2 liter fermentor (22).

degradation of enzyme, resulting in an 85% loss of total activity within 4 h, was observed just prior to the onset of the stationary phase (Figure 2A). To solve this problem, the factors affecting enzyme production and stability, as well as the nutrients required for bacterial growth were examined. By using a defined fermentation medium consisting of glucose (8 g/l), heparin (1g/l), sulfate (2 x 10-3M), salts (10-4M), and amino acids (histidine and methionine, 0.2 g/l), the enzyme titers were increased to 96,000 units per liter of fermentation broth, and the volumetric productivities were increased to 1,0480 units/l (Figure 2B) (18). The enzyme preparation was also stable

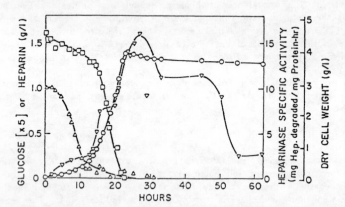

Figure 2B. Results of a typical fermentation in defined protein digest medium showing glucose (□), heparin (△), heparinase specific activity (▽), and dry cell weight (O) as a function of time in a 2 liter fermentor (22).

87

into the stationary phase (Figure 2B).

Recently, in a study of the metabolic regulation of the enzyme synthesis, we found that heparinase synthesis was repressed by the presence of sulfur-containing materials such as sulfate and sulfite (19). By the use of the same fermentation medium containing a low level of sulfate (4.5×10^{-5}M), heparinase was produced in the absence of the previously required inducer, heparin (Figure 3) (15). Although this method gave a

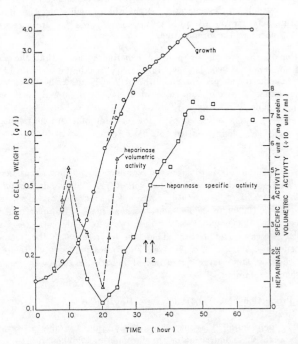

Figure 3. Results of a typical fermentation in low sulfate medium showing heparinase specific activity (□), and dry cell weight (O) as a function of time in a 2 liter fermentor (15).

slightly lower specific activity of heparinase than that of the high sulfate medium, it significantly reduced the cost of fermentation since it obviated the need of heparin as the inducer.

HEPARINASE PURIFICATION

Our goals in the heparinase purification were twofold: (A) to purify the enzyme to homogeneity for the characterization of heparinase; and (B) to produce a large quantity of catalytically pure heparinase for the blood deheparinization studies.

(A) Purification and Characterization of Homogeneous Heparinase

Heparinase was purified to homogeneity by several successive chromatographic procedures including hydroxylapatite ion exchange, Sephadex G-100 superfine gel filtration, chromatofocusing with PBE 94 and Polybuffer 96-CH_3COOH, and Sephadex G-100 superfine gel filtration (14). The homogeneity of the purified heparinase was demonstrated as a single band shown on both the SDS polyacrylamide gel electrophoretic system which separates polypeptides on the basis of molecular weight and the acid-urea gel electrophoretic system which separates polypeptides mainly on the basis of charge (Figure 4). Densitometer traces of both gels also indicated the presence of a single polypeptide (>99% pure) in the preparation.

Characterization of the completely purified heparinase shows that the enzyme has a molecular weight of about 43,000 daltons, an activity maximum at pH 6.5 and 0.1 M NaCl, a stability maximum at pH 7.0 and 0.15 M NaCl, an Arrhenius activation energy of 6.3 Kcal/mole, and a Km and Vm value of 8.04 x 10-6 M and (9.85 x 105 M/min, respectively (14). Amino acid analysis (Table I)
shows that the enzyme is characterized by a low content of sulfur-containing amino acids (< 2%) and a relatively high content of basic amino acids (14%) such as lysine, arginine, and histidine (14).

Although it would be ideal to use the completely purified heparinase for blood deheparinization studies, the total activity recovery (<1%) after purification was too low to satisfy the high demand of heparinase required for such studies (a heparinase reactor would normally require 10,000 to 20,000 units of heparinase. However, since all proteins were immobilized onto the reactor, only those proteins with catalytic activities were crucial, impurities which would either affect the efficiency of the reactor or cause untoward side effects. Our next objective in purification was therefore to develop methods which would produce large quantities of heparinase that was free of all the of her catalytic enzyme contaminants.

(B) Large Scale Production of Catalytically Pure Heparinase

In addition to heparinase, Flavobacterium heparinum contains a variety of other catalytic enzymes such as chondroitinases, hyaluronidase, and heparitinase which act on heparin-like glycosaminoglycans, as well as sulfatases and glycuronidases which act on the heparin degradation products (14,17). Based on the difference in the isoelectric points of these enzymes, we developed a method which would separate heparinase from all the other catalytic contaminants (21). The method includes batch procedures of hydroxylapatite chromatography and a "negative" adsorption (i.e. to adsorb out contaminants) by QAE- Sephadex at pH 8.3. Hydroxylapatite separates heparinase from contaminants of chondroitinases, hyaluronidase, and glycuronidases, while QAE-Sephadex further isolates heparinase from the other contaminants including heparitinase and sulfatases. The resulting heparinase preparation contains less than 1% of each of the contaminating enzymes except for heparitinase which is present at a maximum level of 4% (21). It also shows identical properties and kinetics to those of the completely purified heparinase (21). More important, the method is conducive to producing gram quantities of catalytically pure heparinase in 4-5 h (14,21).

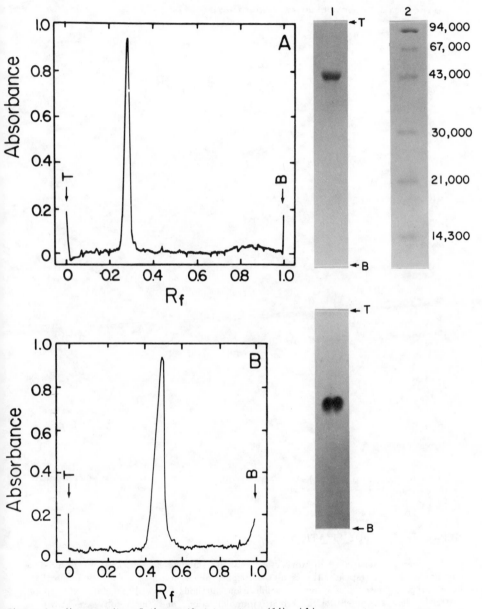

Figure 4. Homogeneity of the purified heparinase (14). (A)
Densitometry trace of the SDS-polyacrylamide gel (slot 1 on the right); slot 2, markers.
(B) Densitometry trace of the acid-urea polyacrylamide gel (right). T, top of the gel; B,
bottom of the gel. R represents the mobility of the polypeptide band.

Amino Acid Composition of Heparinase (14).

Amino acid	Moles (%)	Nearest integer[a]
Half-cystine	0.89	4
Aspartic acid	12.58	52
Threonine	5.59	23
Serine	6.17	25
Glutamic acid	10.12	25
Proline	5.23	21
Glycine	8.62	35
Alanine	12.21	50
Valine	3.68	15
Methionine	0.74	3
Isolucine	3.87	16
Leucine	6.93	28
Tyrosine	3.93	16
Phenylalanine	5.01	20
Lysine	9.04	37
Histidine	2.65	11
Arginine	2.50	10
Tryptophane	ND[b]	ND[b]

[a]The value are calculated assuming that the molecular weight of heparinase is 43,000 daltons.

[b]ND, not determined.

HEPARINASE IMMOBILIZATION

In order for heparinase to work effectively in removing heparin and not to produce any immunological response, the catalytically pure heparinase was covalently immobilized to a support material. Many immobilization methods as well as different support materials were examined for heparinase immobilization, and the results are shown in Table II. The supports giving the highest level of immobilizing heparinase activity include Sepharose 4B (78 - 91%) and Sephadex (5 - 48%) activated by cyanogen bromide (CNBr). It appears that macroporous supports, possibly because of their large surface area, increase the level of total activity that can be recovered on immobilization. Examination of the Sephadex G series suggests that an increase in pore size results in an increase of the immobilized activity even when a large amount of protein is immobilized. The reduced activity observed may be due to the electrostatic repulsion between the support and the highly negatively charged substrate, heparin.

The support material that yielded the highest activity retention was Sepharose 4B. However, because of the low mechanical strength, Sepharose 4B was unable to withstand the high operating flow rate (250 cc/min) generally required by an extracorporeal blood circulation procedure. Beads were found to fracture and pass through the reactor when the reactor was perfused with blood at a flow rate of 250 cc/min. To solve this problem we strengthened the beads by increasing the percentage of agarose in the gel, and by crosslinking the beads with 2,3 dibromopropanol (15). No bead fracturing was observed for the crosslinked 8% agarose gel when it was tested against blood at a flow rate of 250 cc/min (15). Heparinase was also found to be successfully immobilized onto the CNBr-activated 8% crosslinked beads with a high retention of activity (>60%) (15).

Heparinase Immobilization on Various Support Materials (22.23)

Support Material	Coupling Reagent	Activity Recovery (%)
Sepharose 4B	CNBr	78 - 91
Sephadex G-200	CNBr	48
G-150	CNBr	47
G-100	CNBr	40
G-75	CNBr	33
G-50	CNBr	9
G-15	CNBr	5
PHEMA	CNBr	1.6
PHEMA (Spheron)	CNBr	0.03
PMA-CO_2H	EDC	0.01
PMA-$CONH(CH_2)_2OH$	CNBr	0.00
PMA-$(CH_2)_2NH_2$	$(CH_2)_6(CHO)_2$	0.00
PMA-$(CH_2)_6NH_2$	$(CH_2)_6(CHO)_2$	0.00
Dacron	$(EtO)_3Si(Ch_2)_3N=CH(CH_2)_3CHO$	0.20
Dacron-NH_2	$(CH_2)_6(CNO)_2$	0.00
Dacron-CO_2H	Woodwards K Reagent	0.20
Silicone-NH_2	$(CH_2)_6(CHO)_2$	0.00

Abbreviation: CNBr, cyanogen bromide; PHEMA, poly(2-hydroxyethyl methacrylate; PMA, polymethyl methacrylate; EDC, 1-ethyl-3-(3-dimethylaminopropyl) carbodiimide.

Preparation of the Heparinase Reactor.

A fluidized bed heparinase reactor was constructed from a Bentley AF-1025 arterial blood filter (15), and was loaded with crosslinked 8% agarose beads containing immobilized heparinase.

In Vivo Testing of the Heparinase Reactor

For the in vivo experiments, blood access is provided by a Thomas shunt installed in the carotid artery and jugular vein. From the animal, the blood passes through the arterial tubing, an artificial kidney, a venous tubing line with a thrombus trap, the reactor, and back to the animal through a venous tubing equipped with another thrombus trap. Heparin was given intravenously prior to the experiment, and was also continuously infused into the animal during the experiment. Blood samples were drawn from two sampling ports located before and after the reactor. The single pass conversion was determined by measuring the blood clotting time (APTT assay) in samples drawn before and immediately after the reactor.

Previous results have shown that a heparinase reactor can remove over 90% of heparin's anticoagulant activity in dogs within minutes (24,25). However, because of the smaller size and lower blood flow rate of dogs compared to humans, reactors were run at about 50 cc/min which was considerably less than the 250 /min expected in the clinical situations of renal dialysis or pediatric blood oxygenation. In order to test the reactor in a manner as close as possible to a real clinical kidney dialysis procedure, sheep were selected as the animal model. Sheep have comparable blood volumes and flow rates to human beings, and have been widely used for artificial heart implantation studies (26). Three active reactors containing immobilized heparinase (7,000 to 12,000 units) were tested for efficiency, hemocompatibilty and safety. The testing procedures were performed for a period of 1 h since the animal became less manageable for longer periods of time. Preliminary results show that the single pass conversion has varied between 30 - 70%, depending upon the total heparinase activity in the reactor. No hemolysis was observed for any of these reactors over the 1 hour period. The red blood cell counts remained at their initial value throughout the entire time. The white blood cell counts dropped to 50% of their initial value within the first 20 min of the treatment, and returned to normal by the end of the experiments. The platelet counts dropped 40 to 50% of their initial value for the most active reactor, and to 60 - 80% for the least active reactor. The drop in white blood cell counts and platelet counts is well within the range of a normal kidney dialysis procedure.

Toxicology Studies

A number of toxicology studies have been conducted. In the studies with dogs in which the blood was exposed to the immobilized heparinase filter for 90 min, blood was taken from the dogs one, two, and five months after the experiments and no antibodies to heparinase were detected (24). In addition, the heparin degradation products, which were prepared by a complete digestion of heparin with heparinase, were tested for cytotoxic and mutagenic effects on Salmonella typhimorium. No cytotoxicity or mutagenicity was observed even with concentrations 1000 times in excess of those which would be anticipated clinically (24).

93

Studies were also conducted to examine the rate at which heparin degradation products were eliminated from the body. Using both anticoagulant tests and radiolabelled compounds, heparin degradation products were found to be excreted at a much more rapid rate than parent heparin in normal rats and in nephrectomized rats (nephrectomized rats were examined because kidney dialysis was one of the targets for the reactor) (27).

SUMMARY

The development of an immobilized heparinase reactor was discussed. The fermentation of Flavobacterium heparinum has been approved to provide an inexpensive source of large quantities of heparinase. The enzyme has been purified to homogeneity and characterized with regard to its properties and kinetics. A method which is suitable for large scale production of catalytically pure heparinase has also been developed in order to support the high demand of the enzyme for blood deheparinization studies. The catalytically pure enzyme has been immobilized to crosslinked 8% agarose beads and used to construct a heparinase reactor. The reactor has been tested in vivo for efficiency, safety, and hemocompatibility. Products of the enzymatically graded heparin have been tested for cytoxicity and mutagenecity. This heparinase reactor may be a prototype for other enzymatically catalyzed removal systems.

References

1.   Leonard, A. and Shapiro, F.L.;Ann.Intern.Med.; 82: 650, 1975.

2.   Kelton, J.G. and Hirsh, J.;Semin.Hematol.; 17: 375, 1980.

3.   Lazarus, J.M.;Kidney Int.; 18: 783, 1981.

4. Galen, M.A., Steinberg, S.M. and Lourie, E.G.;Ann.Intern.Med.; 82: 359, 1975.

5.   Milutinovich, J., Follette, N.C. and Scribner, B.H.;Ann.Intern.Med.; 86: 189,1977.

6.   Fletcher, J.R., McKee, A.E., Mills, M., Suyden, K.C. and Herman, C.M.;Surgery; 80: 214,1976.

7.   Smith, M.C., Kowit, D., Crow, J.W., Cato, A.E., Park, G.D., Hassid, A. and Dunn, M.J.; Am.J.Med.; 73: 669, 1982.

8.   Laned, A., McGregor, J.R., Michalski, R. and Kakkar, V.V.;Thromb.Res.; 12: 256, 1978.

9.   Choay, J., Lormeau, J.C., Petitou, M., Sinay, P., Casu, B., Oreste, P., Torri, G. and Gatti, G.; Thromb.Res.; 18: 573, 1980.

10.   Miyama, H., Harumiya, N., Mori, Y. and Tanzawa, H.;J.Biomed.Mater.Res.; 11: 51, 1977.

11.   Olsson, P., Larsson, R. and Radergran, K.; Thrombos.Haemostas.(Stuttg.); 37: 274, 1977.

12.   Jaques, L.B.;Can.Med.Assoc.J.; 108: 1291, 1973.

13.   Lowry, O.H., Rosenbrough, N.J., Farr, A.L. and Randall, R.J.;J.Biol.Chem.; 193: 265, 1951.

14.   Yang, V.C., Linhardt, R.J., Bernstein, H., Cooney, C.L. and Langer, R.;J.Biol.Chem.; 260: 1849, 1985.

15.   Bernstein, H., Yang, V.C., Cooney, C.K. and Langer, R.;Methods Enzymol.;In press.

16.   Yang, V.C., Bernstein, H., Cooney, C.L., Kadam, J.C. and Langer, R.;Thromb.Res.;Submitted, 1980.

17.   Linker, A. and Hovingh, P.;Methods Enzymol.; 28: 902, 1972.

18.   Galliher, P.M., Cooney, C.L., Langer, R. and Linhardt, R.J.;Appl.Envir.Micro.; 41: 360, 1981.

19.   Galliher, P.M., Lindhardt, R.J., Conway, L.J. Langer, R. and Cooney, C.L.;Eur.J.Appl.Mircrobiol. Biotechnol.; 15: 252, 1982.

20.   Yang, V.C., Morgan, L., McCarthy, M.T. and Langer, R.;Carbohydr.Res.; 143: 294, 1985.

21.   Yang, V.C., Bernstein, H., Morgan, L. and Langer, R.;Biochem.J.;Submitted 1986.

22.   Langer, R., Linhardt, R.J., Klein, M., Galliher, P.M., Cooney, C.L. and Flanagan, M.M.;In:biomaterials: interfacial phenomenon and applications, advance in chemistry sypmosium series.(Cooper, S. and Peppas, N. eds.); American Chemical Society, Washington D.C.; 199: 493, 1982.

23.   Linhardt, R.J., Cooney, C.L., Tapper, D., , Zannetos, C.A., Larsen, A.K. and Langer, R.; Appl.Biochem.Biotechnol.; 9: 41, 1984.

24.   Langer, R., Linhardt, R.J., Hoffberg, S., Larsen, A.K., Cooney, C.L., Tapper, D. and Klein, M.;Science; 217: 261, 1982.

25.   Langer, R., Bernstein, H., Larsen, A., Yang, V., Tapper, D. and Lund, D.; Am.Soc.Artif.Intern.Organs.J.; 8: 213, 1985.

26.   Murray, M.K. and Olsen, D.B.; Am.Soc.Artif.Intern.Organs.J.; 8: 128, 1985.

27.   Larsen, A., Hetelekidis, S. and Langer, R.; Pharmacol.Exp.Ther.; 231: 373, 1984.

PRODUCTION OF BIOMOLECULES BY IMMOBILIZED ANIMAL CELLS

Kjell Nilsson
Pure and Applied Biochemistry
University of Lund
POB 124
S-221 99 Lund
Sweden

ABSTRACT

The high value of biomolecules produced by animal cells has created a considerable interest in systems for their cultivation. However, factors like complex media, sensitive cells, anchorage-dependency and relatively slow growth rate have resulted in only limited success for large scale culture.

In order to improve cell stability and enhance the production, I have developed methods for immobilization of these cells, figure 1, suspension-grown cells have been entrapped in agarose beads while anchorage-dependent cells have been adsorbed on gelatin microcarriers. A further development has resulted in a semientrapment method for anchorage-dependent cells on/in macroporous gelatin beads.

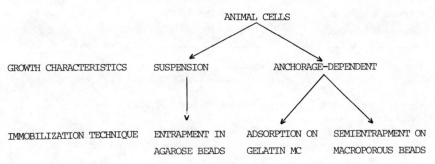

Figure 1. Preparation of immobilized animal cells.

95

Entrapment of Animal Cells in Beaded Agarose

The polysaccharid agarose has several properties which makes it useful for the entrapment of animal cells. Its porous structure allows free diffusion of macromolecules with molecular weights of more than 10^6 daltons.

Partitioning of charged molecules is avoided due to its neutral matrix. The property which makes agarose suitable for cell immobilization is its marked hysteresis. A much higher temperature is thus required for solubilization then for gel formation. A suitable quality for cell immobilization is Sigma Type VII, which is produced by a chemical modification of natural occurring agarose. This quality is thus soluble at 37 °C and has a gel forming temperature below 30 °C. Agarose is also compatible with the majority of suspension cells and is frequently used for cloning of these cells. Previous attempts of using agarose as a matrix for the entrapment of cells has resulted in a homogeneous block which had to be fragmented in order to obtain a desirable size of the biocatalyst. This fragmentation results in loss of cells, irregularly formed particles with a decreased mechanical strength and usually a broad size distribution, the latter leads to poor flow characteristics in packed bed reactors. These disadvantages are avoided if the biocatalyst is prepared in a spherical shape with a uniform size. I have therefore developed a two-phase method for the entrapment of cells in beaded agarose(1). The method is based on a non-toxic hydrophobic phase in which the mixture of cells and agarose is dispersed. The hereby produced droplets of agarose are subsequently solidified by cooling and after removal of hydrophobic phase the immobilized biocatalyst is ready for use. The ideal phase should be non-toxic, inexpensive and easy to separate from the water phase. A hydrophobic phase which fulfills all these requirements is paraffin oil. The method has been used for the entrapment of a number of animal cells, Table I

Table I. Production of biomolecules by agarose-entrapped animal cells.

Cell	Biomolecule	Reference
Mouse/mouse hybridoma	IgG	1
Gibbon lymphoblastoid	Il 2	1
Mouse/mouse hybridoma	IgM	2
Human/mouse hybridoma	IgG	3

The cells listed in Table I have been entrapped in different concentrations of agarose (12.5%, w/v) and at different cell concentrations (3.10^6-20.10^6 - 20.10^6 cells/ml beads).

The immobilized cells have been cultured in roller bottles, spinner bottles and draft tube fermentors in volumes ranging from 50 ml up to 10 litres. Due to the bead size (0.1 - 0.3 mm) continuous operation is easily performed and as the cells are retained inside the beads, a product stream free from contaminating cells is obtained.

However, the formed network is not suitable for the attachment of anchorage-dependent cells. These cells are preferable when immobilized by adsorption on the surface of small beads, microcarriers.

Gelatin Microcarriers

Anchorage-dependent cells require a surface for attachment and subsequent growth. Production of a large number of cells can only thus be obtained if a large surface area is provided. An ingenious solution to the problem of providing a large surface area in a small volume was reported by van Wezel in 1967(4). He was able to culture anchorage-dependent cells on the surface of small beads of an ion-exchanger (DEAE-Sephadex). The small bead size (0.2 mm) resulted in a surface area of 300 cm^2/ml beads. However, when used in high concentrations these beads were found to be toxic for the cells. This toxicity was manifested by the failure of many cells to survive the early stages of culture, long lag periods and limited cell yields at the plateau stage of the culture. It was not until 1979 that Levine and co-workers (5) found that this toxicity could be eliminated through a decrease of the bead charge. This optimerized microcarrier showed no tendency to destroy the inoculum, caused no significant lag phase, and allowed cell growth up to high concentrations.

By using this approach, a number of optimerized microcarriers have been prepared from different synthetic materials. These materials include dextran, poly-acrylamide, poly-styren, tris-acryl and cellulose.

Another type of supporting material for culturing of cells is represented of collagen and its denatured form, gelatin. Collagan would seem to be, at least functionally, the ideal choice since it is the natural substrate for cells. In order to mimic cell attachment in vivo a microcarrier based on gelatin was prepared (6). Gelatin was used as the cell attachment and is equally good or even better than collagen. Cheap raw material and a high solubility in water are further advantages. The microcarrier exhibits no signs of toxicity, such as loss of innoculum and long lag periods.

A particular interesting property of this microcarrier is the superior procedure for cell release. In contrast to the normally applied trypsin treatment, which involves destruction of the cell surface to accomplish cell removal, the gelatin matrix is dissolved by collagenases thereby releasing the cells. This leaves the cells intact and viable which, besides a higher yield, is an advantage when the cells are used for immunization or when analysis of surface antigens will be done.

Human immune interferon and human tissue plasminogen activator have been produced by the culture of recombinant CHO-cells on these microcarriers. In the case of plaminogen activator a much higher production was obtained with gelatin beads as compared to dextran based microcarrier (7). These gelatin microcarriers are commercially available from KC-Biologicals (trade name Gelibeads).

Macroporous Gelatin Microcarriers

Problems encountered with the large scale culture of anchorage-dependent cells are cell yield and the cells fragility. With the gelatin microcarrier, high yield cells are easily obtained as the cells are released from the beads with retained viability. However, as the cells are located on the surface they are subjected to mechanical stress from the mixing system. Another problem is the inherent contradiction in the choice of bead size which results in a suboptimal system. As the cells usually are not transferred from one bead to another, each bead has to be innoculated with a minimum number of cells (5-10 cells/bead). As the bead size decreases, the surface area increases but at the same time the number of cell divisions which can be obtained on each bead decreases and the required size of innoculum increases. Thus, a small bead size results in a large number of discrete steps in order to reach the final protection volume.

If the bead size increases, the surface area is decreased but the required inoculum size is decreased and a higher number of cell divisions are obtained on each bead. Also in this case, a large number of discrete steps is required before the final production volume is reached. This contradiction can be solved if, besides the surface area, also interior volume of the beads is available for cell growth. The required inoculum size is thus decreased by using a lage bead size which also results in a large number of cell divisions. At the same time a large surface area available for cell growth is provided in the interior of the beads. As a result the number of culturing steps are decreased. This new type of microcarrier may be prepared from different materials such as dextran, poly-acrylamide, etc. (8). However, due to the advantages of gelatin as a substrate for cell attachment and growth, it has been used as the matrix. The microcarrier is prepared by a continuous addition of a hydrophobic phase to a gelatin solution (9). In the first stage, the added hydrophobic phase will form droplets in the gelatin solution. Further addition results in saturation of the gelatin solution with droplets of the hydrophobic phase. Still further addition will result in a phase inversion thereby forming droplets of gelatin in the hydrophobic phase. If the droplets are stabilized with emulsifiers, the gelatin droplets will be saturated with droplets of the hydrophobic phase. By cooling the dispersion the gelatin droplets are solidified and after removal of the hydrophobic phase the gelatin matrix is crosslinked with glutaraldehyde to enhance the mechanical strength. The resulting beads, figure 2, are highly porous.

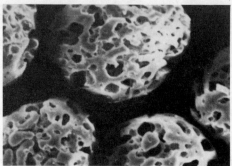

Figure 2.   SEM-picture of a macroporous gelatin bead.   Bar size 0.1 mm.

When tested for their ability to support cell growth for two established cell lines, Vero and BHK 21, it was found that about twice as high cell yields were obtained as compared to solid beads. Due to the cavities a higher number of cells attached to the porous beads. This is probably due to the more favourable conditions which are obtained at the surface. Another advantage resulting from the porosity is a low density which reduces the power input necessary for keeping the beads suspended.

REFERENCES

1. K. Nilsson, W. Scheirer, O. W. Merten, L. Ostberg, E. Liehl, H. W. D. Katinger and K. Mosbach. Entrapment of animal cells for the production of monoclonal antibodies and other biomolecules. Nature 302, 629-630 (1983)

2. W. Scheirer, K. Nilsson, O. W. Merten, H. W. D. Katinger and K. Mosbach. Entrapment of animal cells for the production of biomolecules such as monoclonal antibodies. Develop. biol. Standard. 55, 155-161 (1984)

3. K. Nilsson, W. Scheirer, H. W. D. Katinger and K. Mosbach. Production of monoclonal antibodies by agarose-entrapped hybridoma cells. Methods in Enzymology 121,352-360 (1986)

4. A. L. van Wezel. Growth of cell-strains and primary cells on microcarriers in homogeneous culture. Nature 216, 64-65 (1967)

5. D. W. Levine, D. I. C. Wang and W. G. Thilly. Optimization of growth surface parameters in microcarrier cell culture. Biotechnol. Bioeng. 21, 821-845 (1979)

6. K. Nilsson and K. Mosbach. Preparation of immobilized animal cells. FEBS Lett. 118, 145-159 (1980)

7. K. Nilsson and K. Mosbach. Immobilized animal cells. Develop. Biol. Standard. 61 in press

8. K. Nilsson and K. Mosbach. Macroporous particles. Swedish patent application no 8504764-5

9. K. Nilsson, F. Buzsaky and K. Mosbach. Growth of anchorage-dependent cells on macroporous microcarriers. Bio/Technology in press

UREA HYDROLYSIS STUDIES ON MICROBIAL CELLS BOUND TO CHITIN WITH GLUTARALDEHYDE AS COMPARED TO TOLUENE DIISOCYANATE

John M. Melnyk
St. Bonaventure University
Olean, N. Y.

ABSTRACT

Cells of Proteus vulgaris possessing high levels of urease activity were bound to chitin using glutaraldehyde and toluene diisocyanate. Following immobilization, the cells ability to hydrolyze urea was evaluated and compared. Toluene diisocyanate was shown to be the better binding agent. The activity obtained was dependant on the amount of cells loaded on the chitin, the optimum cell to carrier ratio being 1;1. The optimum pH of the reaction was unchanged from that of the cell suspension. The immobilized cells were placed in a continuous reactor and the effect of flow rate and temperature were studied. Finally, the operational half-life of the immobilized cells was studied.

KEYWORDS

Chitin, Immobilized Cells, Immobilization, Immobilized Cells, Proteus vulgaris.

INTRODUCTION

In 1975, Stanley et al (1), became the first to report using chitin for the immobilization of enzymes. While chitin had been used for some time as an adsorbent for enzyme purification (2), it had never been used with the intention of attaching the enzyme for reuse.

Native chitin is found in the exoskeleton of marine crustaceans. It is composed of roughly equal parts of calcium carbonate and chitin. Chitin is a cellulose-like polymer of 2-amino-2-deoxy-D-glucose units held together by β(1-4) linkages. It has attracted attention because it had been reported that the amino groups are roughly 15% deacylated (1). This led the Stanley group and others (3, 4, 5, 6) to the conclusion that enzymes might be covalently bound to those groups with the bifunctional cross-linking agent glutaraldehyde.

Urease was bound to chitin by Iyengar and Rao (6) using glutaraldehyde. They felt that the immobilized enzyme would be useful in treating urea-rich waste effluent from

101

fertilizer plants.

It seemed that application, or possibly some other (like analysis) might be facilitated by immobilizing cells containing high levels of urease activity. This would save the step of purifying the enzyme before immobilization and would also increase the operational half-life of the enzyme.

EXPERIMENTAL

Chitin Preparation

Chitin from the snow crab (Chionoecetes bairdi) was obtained from the Bio-Dry Corp., Albany Oregon as crushed flakes. It was ground further to a particle size of between 45 and 125 mesh. The powder was then moistened with water and treated with 6N hydrochloric acid to remove the calcium carbonate. When no more carbon dioxide was produced the mixture was placed under reduced pressure. The pressure was released and the procedure was repeated several times to improve penetration of acid into the chitin interior. The chitin was washed several times with water until pH paper showed a neutral pH when pressed lightly on the surface.

A second method was studied which employed the additional step of a warm alkali bath (3 hr. with 5N potassium hydroxide). Again, the chitin was washed until the pH was neutral.

Organism

Cells of the genus Proteus were chosen for study since they are known to have high levels of urease activity. Proteus vulgaris (ATCC 13315) were grown on Tryptic Soy Broth (TSB) and maintained on Tryptic Soy Agar (TSA).

Cell suspensions were prepared by centrifuging samples of TSB after 18 hr. of growth and resuspending them in 0.1 M phosphate buffer (pH 7.0) to an OD of 1.825.

Cell pastes were prepared by centrifuging 18 hr. cultures at 5000 rpm for 5 minutes at $2\,^{\circ}$C.

Assay

Cell urease activity was determined by estimating the amount of ammonia formed. One ml. samples of cell suspensions were added to 1.0 ml. samples of urea (0.18M in 0.1M phosphate buffer, pH 7.0) and shaken for 15 minutes at $22\,^{\circ}$C. ($\pm$ $1\,^{\circ}$C.). The reaction was stopped by chilling the mixture in a $2\,^{\circ}$C. water bath. An aliquot (1.0 ml) was tested for ammonia present using Nesslers reagent (7).

The immobilized cell urease was tested by reacting 10.0 ml of urea (0.18M in 0.1M phosphate buffer, pH 7.0) for 30 minutes by shaking at $22\,^{\circ}$C. The reaction was stopped by filtering the chitin from the solution. A 1.0 ml aliquot of the solution was tested as above.

The feed solution for the continuous reactor was testd for urea concentration using a modified p-dimethylaminobenzaldehyde method of Ghose and Kanna (8). This assay was modified to use a 1.0 ml sample size.

Optimum pH

The optimum pH for both the cell urease in the suspensions and in the immobilized cells were determined by the previously described assays. The urea solutions were prepared in 0.1M phosphate buffers at pH's ranging from 5.7 to 8.2. Fresh immobilized cell samples and suspensions were used for each test.

Immobilization Procedures

Cells were bound to the chitin by covalent attachment using glutaraldehyde and toluene diisocyanate (TDI). Following immobilization, samples of immobilized cells prepared by the two methods were tested for activity and compared.

Ten gram samples of the chitin, prepared as above, were pretreated with glutaraldehyde by the addition of a 1.0% solution. The mixture was shaken gently for about 30 minutes and stored for 24 hours at 2°C. At the end of that time, the glutaraldehyde was washed repeatedly with water. A 10.0 gram sample of the cell paste was diluted in 50 ml of 0.1M phosphated buffer (pH 6.7) and added to the pretreated chitin. It was shaken gently for about 30 minutes and stored at 2°C. for 48 hours. The mixture was filtered and washed with water until the effluent was clear.

The same procedure was followed with the TDI pretreatment except that the TDi was a 2% solution prepared in 95% ethanol. This necessitated a different wash procedure following cold storage. Excess TDI was removed by a series of washes in decreasing concentrations of ethanol, ending in several water washes.

Loading Dose

The effect of the amount of bacteria immobilized per gram of chitin was studied by varying the weight of the cell paste added to the chitin.

Continuous Reactor

Chitin which contained cells immobilized with TDI was placed in a fluidized bed reactor (Figure 1). The enzymatic reaction could then be studied over a range of conditions and a long period of time.

The temperature was maintained within $\pm$ 1°C. by a heat controlled circulating water bath. The urea was delivered using a variable speed peristaltic pump and the reaction mixture was kept in motion by a variable speed magnetic stirrer.

Figure 1. Continuous Reactor

Optimum Temperature

The optimum temperature of the enzyme reaction was studied by placing a 10.0 gram sample of the immoblized cell-chitin in the reactor and varying the temperature. Aliquots of the reaction mixture were asceptically removed and assayed as described earlier.

RESULTS

Chitin Pretreatments

Chitin is an inexpensive, strong glucoseamine polymeric material, which is of interest for the immobilization of enzymes and cells because of the free amine groups found on its surface. Other researchers (1, 6) have studied the effectiveness of some pretreatments on releasing an increased number of amines for binding.

In this work, two pretreatments were studied; an acid treatment, which is almost standard, for removing calcium carbonate from the surface. The other is an additional alkali treatment with potassium hydroxide. Table 1 gives the results of the effects these treatments have on cell urease bound by TDI.

Treatment	Activity (ug Ammonia/ml/min)
Hydrochloric Acid	45.0
Hydrochloric Ac/Potassium Hydroxide	11.7

Table 1.   Immoblized Enzyme Activity Following Acid and Alkali Pretreatments.

This study indicates that the alkali bath does not have a beneficial effect on cell immobilization.

The second major area of interest was the effect of the cross-linking agent on the amount of activity obtained by immoblization. Table 2 is a comparison of the amount of enzyme activity which results from immoblization with 1% solutions of glutaraldehyde and TDI.

	Activity (ug Ammonia/ml/min)
Glutaraldehyde	12.0
TDI	46.7

Table 2. Glutaraldehyde vs. TDI as Binding Agent

In addition to this study, it was felt that consideration should be given to the effect that a variation in the concentration of TDI would have on the amount of immoblized cell activity obtained. Table 3 gives the results from this study.

TDI %	Activity (ug Ammonia/ml/min)
2.00	30.0
1.00	47.5
0.50	40.0
0.25	41.7
0.10	25.0
0.00	<1.0

Table 3.   Effect of Varying Concentrations of TDI on Immobilized Enzyme Activity.

SEM photographs (Figure 2) show the difference in the number of cells bound with glutaraldehyde vs. TDI.

Figure 2.   SEM Photographs of Glutaraldehyde Bound Cells and Those Bound with TDI.

Optimum pH

Many studies have shown that enzymes immobilized on carriers with active surface groups can have marked changes in their pH behavior after immobilization. Immobilized cells have their enzymes maintained inside the cells and therefore should not show this effect. Figure 3 presents the results of such a study, in which, cell urease was studied before and after immobilization. The peak is broadened with the immobilized cell urease, but the optimum pH remains the same as the cell urease in suspension.

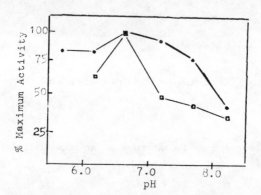

Figure 3.   pH Optima - Immobilized Cells - Cell Suspension

107

Loading Dose

The load of cells added to the chitin was studied by following the immoblization procedure with varying amounts of cells. Table 4 shows the effect of loading on cell urease activity before and after immoblization.

Loading Dose (gr. cells/gr. chitin)	Activity (uf Ammonia/ml/min)
0.00	<1.0
0.10	14.3
0.25	20.0
0.50	41.5
1.00	45.0
2.00	45.0

Table 4. Effect of Loading Dose on Urease Activity.

The data shows the relationship between urease activity and loading dose. There is no increased activity gained after a cell to carrier ratio of 1:1. At that point, all available binding sites or physical space are used so that no additional cells can be attached.

Continuous Reactor

Ten gram samples of the immobilized cells were placed in the reactor and the temperature was varied. The results are presented in Figure 4.

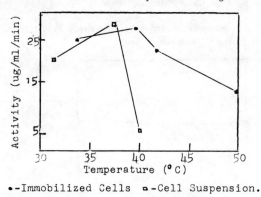

•-Immobilized Cells ◦-Cell Suspension.

Figure 4.   Temperature Optima - Immobilized Cells - Cell Suspension

108

The striking feature of this study is that the immoblized cell urease behaviour versus temperature shows some stability, not seen in the cell suspension. While it is not as dramatic as the stability shown in earlier work (6) with the enzyme, there does seem to be some effect.

The dramatic effects noted with enzymes may be due to the strong bonds which prevent the enzyme from denaturation. In the case of the immobilized cells, the enzyme is in the cell, just as in the suspension but the chitin is now acting as an insulator. This would explain why the optima of the immobilized cells is not shifted very much (only about 3°C.) but instead is broadened.

Another interesting effect that was noted was the return to higher levels of activity, for the immobilized cell urease, following a return to lower temperatures. This would suggest that older or damaged cells were replaced by new cells which were not exposed to the high temperature.

The flow rate of the reactor was studied and the effect on urease activity is presented in Figure 5.

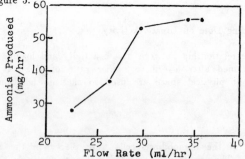

Figure 5. Effect of Flow Rate on Immobilized Cell Urease.

A sample of immobilized cells was studied at 37°C. and a flow rate of 34ml/hr for a ten day period. Figure 6 shows the resulting data.

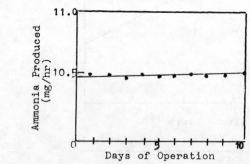

Figure 6. Continuous Operation of Immobilized Cell Reactor

The immobilized cell reactor showed no loss of urease activity over the entire period of operation. There actually seems to be a slight upward bias in the activity curve. In several other trials run under non-ideal conditions, the same pattern of enzyme activity stability is seen for period of up to 35 days.

While this indicates a very long half-life, relative to a study done with immobilized urease (9), a study of longer operation would need to be made, to say conclusively if that is the case.

In an effort to determine the fate of all of the urea entering the system, a study was made of urea levels entering and exiting the reactor. When added to the amount of ammonia produced, 98.5% of the urea was accounted for, indicating that the cells used a small amount as a nitrogen source.

CONCLUSION

Chitin was shown to be useful as a support for the immobilization of cells of Proteus vulgaris. Both TDI and glutaraldehyde attach the cells, however, TDI seems to be the superior agent.

TDI forms stable urethane bonds with hydroxyl groups, which are found in great numbers on the peptidoglycan surface of the gram negative cell. Glutaraldehyde, on the other hand, forms stable bonds with amines by the formation of Schiff's bases. It appears that the hydroxyl groups predominate, thus TDI being the more effective agent.

Immobilization of the cells showed no effect in the pH behavior of the enzyme. Immobilization of enzymes can have consequences on the pH particularly if there are active groups on the surface. The immobilized cell enzyme is not exposed to these effects, and so, shows no pH change.

There was some effect on the enzyme behavior with regard to temperature. The immobilized system showed some slight increase in temperature optimum and a broadening of the range the enzyme would function under. This is most likely due to the insulation effect of the chitin, especially given the fact that ten grams of chitin would have a limited ability to absorb heat. Again, a more dramatic change is seen when the enzyme is immobilized directly (6) because the molecular structure of the enzyme is held tightly in place, and can not denature. This is not seen in immobilized cell urease because the enzyme is basically in its natural environment.

The long-term reactor is the area where immobilization of cells plays a great role. Immobilization, by its nature, gives a much greater degree of control over reactions than conventional batch techniques. This study shows that the possibility exists to further improve, because of the increased operational life of the catalyst. Specifically, an earlier work (9) indicated an operational half-life for immobilized urease of 19 days. This study shows that cells have a much longer half-life. It is possible that given the lower initial activity of the cell reactor (about 13% of the immobilized enzyme reactor), it could be improved by recycling or making a larger reactor to overcome the immobilized enzyme advantage.

110

REFERENCES

1. Stanley, W. L., G. G. Watters, B. Chan and J. M. Mercer, Lactase and other enzymes bound to chitin with glutaraldehyde. Biotechnol. Bioeng. 17:315-326. (1975).

2. Cherkasov, I. A., N. A. Kravchenko and E. D. Kaverzneva, Molec. Biol., 1:41. (1967) (Cited in Stanley 1975).

3. Stanley, W. L., G. G. Watters, S. H. Kelly, B. G. Chan, J. A. Garibaldi and J. E. Schade, Immobilization of glucose isomerase on chitin with glutaraldehyde and by simple adsorption. Biotechnol. Bioeng. 18:439-443. (1976).

4. Stanley, W. L., G. G. Watters, S. H. Kelly and A. C. Olson, Glucoamylase immobilized on chitin with glutaraldehyde. Biotechnol Bioeng. 20:135-140. (1978).

5. Liu, W. H., Enzymatic oxidation of glucose via crab chitin immobilized glucose oxidase and catalase, in Chitin and Chitosan Proceedings of the 2nd International Conference on Chitin and Chitosan, S. Hirano and S. Tokura (Eds.), The Japanese Society of Chitin and Chitosan. (1982).

6. Iyengar, L. and A. V. S. P. Rao, Urease bound to chitin with Glutaraldehyde. Biotechnol. Bioeng. 21:1333-1343. (1979).

7. Chigar, D. A Study of the Metabolism of the Genus Caulobacter. Masters Thesis, St. Bonaventure Univ. (1967).

8. Ghose, T. K. and V. Kannan, Studies on fibre-entrapped whole microbial cells in urea hydrolysis. Enzyme Microb. Technol. 1:47-50. (1979).

9. Melnyk, J. M. A Study of the Urease Activity and Stability of Immobilized Urease and of Immobilized Cells of Proteus vulgaris Using Chitin as the support for both the Enzyme and Cells. Doctoral Thesis, St. Bonaventure Univ. (1986).

BIOFILMS ON ADSORBENT PARTICLES

Graham F. Andrews and Jean Pierre Fonta
Department of Chemical Engineering
State University of New York at Buffalo
Buffalo, New York
14260

ABSTRACT

Biofilms growing on an adsorbent surface like activated carbon have several advantages. Inhibitory substrates and products can adsorb thus reducing their effect on the biomass and previously-adsorbed substrate can desorb and feed the base of the film. These considerations have produced a new design of continuous bioreactor, consisting of a fluidized bed of granular activated carbon stratified based on biomass volume. Heavily-coated particles are removed from the top of the bed, washed to remove excess biomass, treated to remove product, and recycled to the base of the bed. Ethanol production with a flocculent strain of Zymomonas mobilis showed that the bed stratifies as required even with CO_2 bubbles present. Gas slugging problems restricted the inlet concentrations possible in the 1" laboratory reactor, so the production of lactic acid is now being studied.

KEYWORDS

Bioreactor, activated carbon, ethanol, fluidized bed.

INTRODUCTION

Biofilms on Adsorptive Surfaces

Many immobilized-cell reactors contain films of biomass growing on some type of support particle. They include trickling filters used for wastewater treatment, packed-beds proposed for ethanol production[1], and several fluidized-bed designs for anaerobic fermentations[2] and aerobic and anaerobic wastewater treatment[3][4]. Fluidized beds of floc particles such as tower fermenters and sludge-blanket reactors can be included because they represent the limiting case as the size of the support particle goes to zero.

111

All these reactors share the problems associated with the mass transfer resistance in the biomass; substrate must diffuse into the film and product must diffuse out. Since both of these processes require a concentration gradient, the region deep inside the biomass can become a very low substrate/high product environment where metabolic activity is severely inhibited. As far as reactor productivity is concerned, it is just wasted space. Consequently control of the film thickness, either by natural shear, air scouring, natural sloughing or by removing particles from the bed and washing off excess biomass[3], is an important part of the reactor design.

Films growing on an adsorbent support particle such as activated carbon are less affected by these problems. If the adsorbent has been saturated with substrate then, as the film grows, this substrate will slowly desorb and feed the base of the film [5]. Furthermore, any organic products formed can adsorb onto the carbon thus reducing their inhibitory effects on the biomass. The cells themselves are too large to enter the pores of the adsorbent. Indeed, most of the pores in activated carbons are smaller than 50 A, so not even proteins can compete with small organic molecules for the majority of the surface area available for adsorption.

Effectiveness Factors.

In order to quantify the possible gains in reactor productivity to be expected with adsorbent particles the following equation must be solved for the simultaneous diffusion and consumption of a component in the biomass (Fig. 1a).

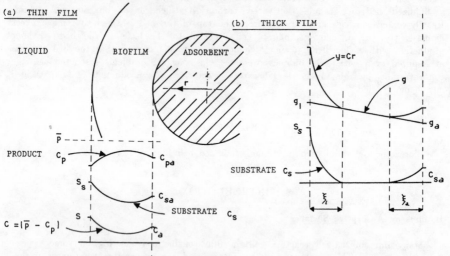

Fig. 1: Concentration profiles in the biofilm

$$\frac{D}{r^2} \frac{d}{dr}(r^2 \frac{dc}{dr}) = \frac{\mu}{Y} \tag{1}$$

$$C = S \quad \text{at} \quad r = R+L$$

$$C = Ca \quad \text{at} \quad r = R.$$

There is one such equation for each component that affects the value of the cell growth rate μ. If the component is an inhibitory product then the right hand side should be negative, but this is avoided by the variable transformation $c = p - C_p$ where C_p is the actual product concentration and p is the concentration that stops growth altogether. (Now $c = 0$ stops growth, just as it would for a substrate). Note that non-growth associated activity and liquid-phase mass transfer resistance have both been ignored in equation [1]. They were included in an earlier analysis [6] but found to be unimportant in the present case.

We assume that the inherent cell kinetics are first order; $\mu = \mu c/K$. This is a reasonable approximation for aerobic waste-water treatment where the substrate concentrations must be kept low in order to prevent the bed from becoming anaerobic, and for many anaerobic fermentations whose kinetics can be described by the linear-inhibition model ($K = p$ in this case). Equation [1] can then be solved for the concentration profile in the biomass.

$$c = \frac{(\theta+\gamma) S \sinh (\bar{r}-\gamma) + \gamma Ca \sinh (\theta+\gamma-\bar{r})}{\bar{r} \sinh \theta} \tag{2}$$

Two effectiveness factors can now be defined:

$$\eta_a = \frac{\text{Rate of diffusion of component from carbon}}{\text{Maximum rate of component consumption if whole particle were biomass exposed to liquid-phase conditions}}$$

$$= - \frac{4\pi R^2 D}{\frac{4}{3}\pi (R+L)^3 \frac{\mu S}{YK}} \frac{dc}{dr}\bigg|_R = \frac{3\gamma}{(\theta+\gamma)^3} \frac{[(\gamma \cosh \theta + \sinh \theta)(Ca/S)-(\theta+\gamma)]}{\sinh \theta} \tag{3}$$

$$\eta_\ell = \frac{\text{Rate of diffusion of component from liquid}}{\text{Maximum rate of component consumption if whole particle were biomass exposed to liquid phase conditions.}}$$

$$= \frac{4\pi (R+L)^2}{\frac{4}{3}\pi (R+L)^3 \frac{\mu S}{YK}} \frac{dc}{dr}\bigg|_{R+L} = \frac{3[(\theta+\gamma)\cosh\theta - \sinh\theta - \gamma Ca/S]}{(\theta+\gamma)^2 \sinh \theta} \tag{4}$$

If the component of interest is a product then "into" and "production" replace "from" and "consumption" in these definitions. However, with the variable transformation given above, nothing else is changed.

The definitions are based on the total particle volume rather than the biomass volume, because this leads to a simpler expression for reactor productivity (defined as mass of the component consumed, or produced, per unit reactor volume per unit time). The total rate of consumption, (or production) of the component per unit volume of solids is $(\eta_\alpha + \eta)\mu S/YK$. So the reactor productivity is the value of $\epsilon_s(\eta_a+\eta)\mu S/YK$ averaged through the reactor. This gives the reasonable result that maximizing the reactor productivity requires a high average solids holdup, ϵ_s, a high average liquid-phase concentration, S, (achieved for fixed inlet and outlet concentrations in a plug-flow reactor), and a high average total effectiveness, $(\eta_a+\eta)$.

The variation of $(\eta_a=\eta)$ with the amount of biofilm on the particle, calculated from equations (3) and (4), is shown in Fig. 2. The quantity x is the ratio of film volume to support particle volume and is given by:

$$1+x = (1+\frac{\Theta}{\gamma})^3 \qquad\qquad (5)$$

Several points are obvious from Fig. 2. First, a monolayer of cells

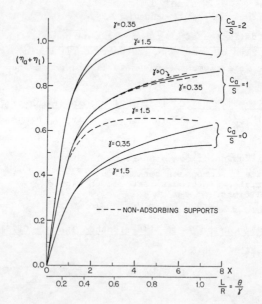

Fig. 2: Total effectiveness factor

would give x of order 10^{-2} which would give negligible effectiveness. Cells that will not form a thick film spontaneously should, therefore, be immobilized not on the surface of solid spheres but inside gel beads.

Second, a high effectiveness requires small support particles. Consider the curves shown for non-adsorbing supports (found by putting $\eta_a=0$ and calculating Ca/S from [3]). For small supports (γ -&rarrow.0) a given x corresponds to a thin biofilm, so there is no mass-transfer resistance in the film and the total effectiveness equals the fraction of the particle volume consisting of biomass; $1/(1+x)$. The result for $\gamma=0.35$ is close to this upper limit (as is the line for and adsorbent support with Ca/S=1, since "no mass transfer resistance in the film" implies Ca=S). The result for $\gamma=1.5$ is significantly lower although this corresponds to a support diameter of only 3 mm for ethanol production and 0.2 mm for aerobic wastewater treatment [7]. It follows that the large supports required to prevent clogging in packed-bed reactors will give very low reactor productivity. Fluidized beds, which can expand to accommodate the volume of growing biomass, have a definite advantage.

Third, adsorbent supports have a definite advantage over non-adsorbing supports if Ca/S is high. Indeed for Ca/S=2 the effectiveness goes over the theoretical maximum of 1. Unfortunately, keeping Ca/S high is impossible if the particle is in a constant environment (in a CSTR for example). A high Ca/S gives a positive η_a because it corresponds to a negative dc/dr_R (equation 3). Substrate is diffusing out of the adsorbent to the film base or an inhibitory product is diffusing in. Neither situation can last forever because the capacity of the adsorbent is finite.

Eventually an equilibrium situation must be established in which there is no diffusion in either direction. Since "no diffusion" implies $dc/dr"_R = 0$, this equilibrium would be identical to the non-adsorbing support case. It follows that the only way to maintain a high Ca/S value is to move the particle continually to regions of lower substrate concentration.

A caution is necessary about equations [3] and [4]. It can happen, depending on the relative concentrations of the different substrates and products in the liquid phase, that the component for which equation (1) is solved is not the limiting component (defined as the one whose concentration reaches zero first in a thick biofilm). For example, glucose may be actually limiting in ethanol production or oxygen in aerobic wastewater treatment. The resulting concentration profiles are shown in Fig. 1 (b). The complete analysis has been given previously [8], but this situation should not be of great practical importance because the inactive zone in the center of a thick film represents wasted space which would not be allowed in a well-designed bioreactor.

The Proposed Reactor

A continuous bioreactor design based on the considerations given above is shown in Fig. 3. Small monosized activated carbon particles coated

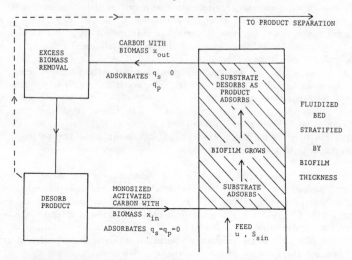

Fig. 3: The proposed reactor

with a thin biofilm are fed to the base of a fluidized bed. Some substrate diffuses through the film and adsorbs on the carbon thus reducing the liquid-phase substrate concentrations, which is a valuable consideration for substrate-inhibited fermentations. Subsequent growth of the biofilm increases the ratio of product to substrate at the film/carbon interface, so substrate tends to desorb and be replaced by product. This desorption provides substrate to the base of the biofilm, thus reducing the effects of the film's resistance to substrate diffusion from the liquid. Growth of the film also (within limits) reduces the settling velocity of the particle. Since liquid-fluidized beds have a strong tendency to stratify, the reduction in settling velocity moves the particle up the bed to regions where the product/substrate ratio is higher in the liquid phase. This further encourages desorption of substrate and adsorption of product, so the top of the bed should contain particles containing only product and coated with a thick film of biomass. They can be removed, washed to remove excess biomass, treated to desorb the product, and returned to the base of the bed.

Operation of the reactor requires an orderly "plug flow" of solids up the bed. This requires three conditions. First the bed must stratify based on the settling velocities of the particles. Any strong solids mixing would destroy this stratification and mean that some particles removed from the top of the bed would contain substrate rather than product. Fortunately, liquid-fluidized beds are known to stratify well [7]. Second, the adsorbent particles must be virtually monosized or the larger particles, having a high settling velocity, will stay near the base of the bed despite the growth of the biofilm while small particles will rise up the bed before the film develops. Third, biofilm growth must reduce the particle settling velocity or the heavily coated particles will be at the bottom of the bed, not the top. This reduction should not be taken for granted. Although biofilm growth reduces the average particle density, which tends to reduce settling velocity, it also increases the diameter, which tends to increase it. The net result, based on a drag coefficient correlation of the form $C_D=f/Re$, is [7].

$$\frac{U_t}{U_{ts}} = \frac{(1+Bx)^{1/(2-e)}}{(1+x)^{1/3}} \qquad (6)$$

This equation is plotted in Fig. 4. For very light support particles

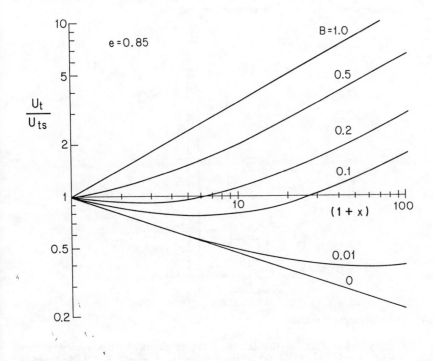

Fig. 4: The effect of biofilm on settling velocity

$(B\longrightarrow1)$ the diameter effect is dominant and film growth always increases U_t. For very dense supports $(B\longrightarrow0)$ the density effect dominates and growth always decreases U_t. Given a wet biomass density of 1.03 g/cm^3 and a wet carbon density of 1.52 g/cm^3, B=.05 so the bed can stratify as required as long as x<8 (approximately). This sets the upper unit on x in the bed. The lower unit (x_{in}) is fixed by the requirement that the average value of $(\eta_a=\eta_l)$ in the bed be high. It is clear from Fig. 2 that this only happens for x>1.

The proposed reactor has most of the attributes required to give high productivity; a high solids holdup; a liquid flow that approximates plug flow; small particles with high effectiveness factors; and a means of reducing the substrate and product concentrations in the liquid phase in order to minimize possible inhibition. A theoretical model of the reactor has been published previously [8]. This paper reports on experience operating a laboratory-scale reactor to produce ethanol from glucose.

EXPERIMENTAL

Experiments were conducted in a plexiglas laboratory reactor 5.8 m in height (Fig. 5). The working section was 2.54 cm in diameter, but the

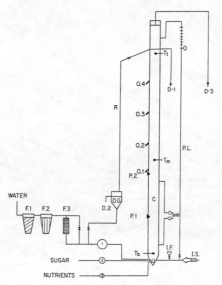

Fig. 5: The experimental reactor

upper 1.05 m was 3.81 cm and the bottom 0.115 m consisted of a conical entrance section. Liquid samples could be taken through septa at points P1 (1.05m) and P2 (2.06m), and four outlet ports (0.1-0.4) were available for solids removal. The solids were re-injected into the base of the bed after treatment using a funnel (1F), syringe (IS) and a system of tubing clamps. The reactor was fed with tap water passed through a particle filter (F1), an activated carbon filter to remove chlorine (F.2) and a 0.2 μm sterilizing filter. The water temperature was raised to 25°C by heating tape wrapped around the inlet tubing and was kept at this level by insulation around the column.

The water was mixed with a concentrated glucose solution from a 50 L carboy and a mixture of nutrients from an 8 L carboy. These carboys were heat sterilized and changed daily. The influent composition was kept constant by pumping the water, sugar and nutrient solutions through three pump heads[1 2 3 in Fig. 5] on a single peristaltic pump. The nutrients in these experiments consisted of 0.1 gram of yeast extract per gram of glucose.

The organism used was a flocullant strain of <u>Zymomonas mobilis</u> (NRRL B12526) and the adsorbent was Purasiv® activated carbon supplied by the Linde Division of Union Carbide Corporation. The particles of this carbon are virtually spherical with an average diameter of 700 μm and a size distribution (measured on a Quantimet Image Analyzer) shown in Fig. 6. It's saturation capacities for glucose

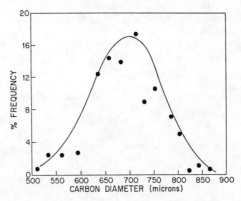

Fig. 6: Particle size distribution

and ethanol, measured in batch isotherm experiments with no yeast extract present, were 0.306 gm/cm^3 and 0.318 gm/cm^3 respectively. The fluidization characteristics of beds containing various (dry) weights of these particles with water in the experimental column is shown in Fig. 7.

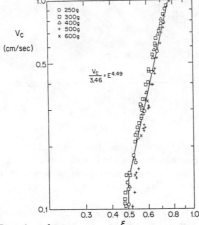

Fig. 7: Fluidization of Purasivcarbon

To start up the reactor the tubing, fittings and activated carbon were steam sterilized. The reactor was then assembled and filled with 0.75 N hydrochloric acid which was recycled around (line R in Fig. 5) for five hours. The reactor was then flushed with sterile water, glucose (50 gm/L) was introduced and recycled for 4 hours to reach adsorptive equilibrium, and finally the yeast extract was introduced and a cell inoculum injected through the septum of port P1. This culture was recirculated slowly around the reactor until a marked opacity indicated that cell growth was complete. The column was then switched to continuous operation, the liquid flow being set to give a bed porosity of roughly 60% in order to minimize agglomeration of the particles.

The bed height, liquid flow rate and liquid-phase concentrations and the inlet, outlet and ports P1 and P2 were measured at regular intervals. Ethanol concentrations were measured by gas chromatography.

RESULTS

A typical set of data are shown in Fig. 8. Significant volumes

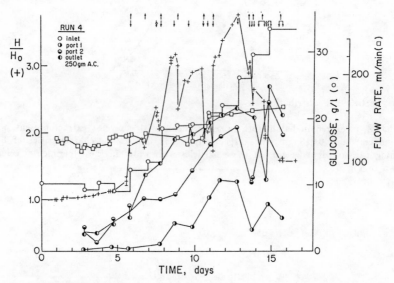

Fig. 8: Results of typical run

of biofilm start to appear after 5 days and the bed expands rapidly to accommodate it. The carbon used in this run had been soaked with 3 N nitric acid for two days prior to being sterilized in an attempt to improve bacterial attachment by oxidizing the surface. This pre-treatment reduced the "lag" period to 5 days from 15 in other runs.

Unfortunately, as the biomass volume and activity increased so did the amount of CO_2 produced. The gas bubbles tended to strip biomass off the particles, forming a bed of floc particles on top of the bed of activated carbon. (These flocs were drained off whenever possible and the H values are those of the activated carbon bed). After 10 days the glucose removal in the bed reached 20 gm/L and the gas production was great enough to form a complete slug across the bed which stripped off most of the biomass, leading to a sharp drop in bed height. This gas slugging made it impossible to reach ethanol concentrations that would be strongly inhibitory to the organism or to reach a steady-state at the high biomass volumes required by the theory. To a first approximation in the absence of gas the average x value in the bed is proportional to $((H/H_o)-1)$ [9], so the values shown in Fig. 8 corrected for the gas holdup give an average x of 1~2, significantly less than the optimum values from Fig. 3. However despite this gas slugging problem, which may or may not happen in a larger-diameter column, several significant observations were made.

The amount of ethanol produced between any two points in the bed while no carbon was being recycled is plotted against the amount of glucose consumed in Fig. 9. The slope gives the yield as 48.5% or 95% of the

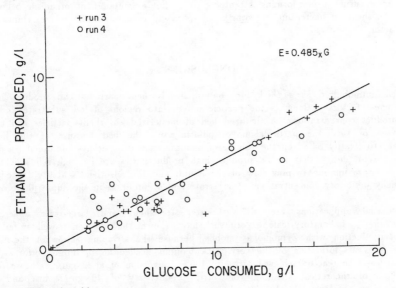

Fig. 9: Ethanol yield

theoretical yield, a reasonable value for this organism.

The reactor could sustain virtually complete conversion of a 20 gm/L glucose feed which gives a volumetric productivity of 40 gm/L hr., an encouragingly high figure in view of the sub-optimal biomass concentration.

The liquid phase does approximate plug flow, as is shown by the wide separation of lines for glucose removal at port 1, port 2 and the reactor outlet.

122

Since activated carbon is black and the Zymomonas biofilm is white the tendency of the bed to stratify based on biomass volume was easily visible. There was indeed a smooth transition from lightly coated, black particles near the bottom of the bed to heavily coated, white particles at the top despite the presence of gas bubbles which are known to promote solids mixing in liquid-fluidized beds. However, experience with large tower fermenters used for beer production [10] has shown that fluidized beds of this scale do stratify with, in this case, large flocs at the base of the bed and small flocs at the top.

Contamination of the reactor by foreign organisms was not a problem during the ethanol production experiments. This was surprising since some runs lasted over a month and the sterilizing filter in the tap water line (F3 in Fig. 5) became clogged after 1-2 weeks and was by-passed. Also no special precautions were taken with the pH (it dropped from 6.5 to 3.5 through the reactor) and rigorous initial sterilization of the plexiglas column was impossible. Scott (2) reports similar experience with a fluidized bed of Zymomonas flocs. During subsequent experiments on lactic acid production in the reactor, using a film-forming Streptococcus strain, a contaminant organism believed to be the sheathed bacterium Crenothrix polyspora did establish itself in large clumps in the reactor.

CONCLUSIONS

Cells immobilized as a thick film on small adsorbent particles can produce high effectiveness factors as long as the particle moves into regions of low substrate and/or high product concentration. A fluidized bed of monosized adsorbent particles can take advantage of this; the particles move uniformly up the bed because of the bed's tendency to stratify based on the volume of biomass on the particle. Additional features of this reactor design that should produce high productivities are its high solids holdup, its close approximation to plug flow of the liquid, and the adsorbent's ability to reduce potentially inhibitory concentrations of substrate and/or products in the liquid phase.

Potential applications are in wastewater treatment and large-scale anaerobic fermentations. Laboratory-scale experiments have shown that it is not well suited to ethanol production with Zymomonas mobilis because the stripping action of the CO_2 bubbles prevents formation of the very thick biofilms needed for optimum performance. Nevertheless, the reactor achieved virtually 100% conversion of a 20 gm/L glucose feed with 95% of theoretical yield, a volumetric productivity of 40 gm/L hr., and no contamination problems. It also demonstrated that the bed stratifies as required even in the presence of gas bubbles. Tests are now proceeding with the homolactic fermentation in order to eliminate problems arising from gas production.

ACKNOWLEDGEMENT

This work was carried out under NSF grant CPE 8204968.

NOMENCLATURE

B :   buoyant biomass density/buoyant density of adsorbent.
C :   concentration in the biofilm.
C_a :   concentration at the film/adsorbent interface
D :   diffusivity in the biofilm.
e :   exponent in drag coefficient correlation.
K :   Monod constant
H :   height of fluidized bed
H_o :   height of fluidized bed of adsorbent at same flow rate.
L :   biofilm thickness.
p :   product concentration that stops growth.
r :   radial distance.
r : $r(\bar{\mu}/DYK)^{1/2}$
R :   radius of adsorbent particle.
S :   liquid-phase concentration.
U_t :   settling velocity of particle.
x :   biomass volume/adsorbent particle volume.
Y :   volume biomass produced/mass of component consumed (produced).
γ : $R\ (\bar{\mu}/DYK)^{1/2}$
η_a :   effectiveness factor for adsorption.
η_l :   effectiveness factor for liquid phase.
ϵ :   bed porosity.
ϵ_s :   solids holdup in reactor.
μ :   specific growth rate.
μ :   maximum specific growth rate.
Θ : $L(\bar{\mu}/DYK)^{1/2}$

LITERATURE CITED

1. O. C. Sutton, J. L. Gaddy, Biotech, Bioeng. <u>22</u> 1735, 1980.

2. C.D. Scott, Biotech. Bioeng. Symp. Series, <u>13</u>, 287, 1983.

3. P. Cooper, B. Atkinson, "Biological Fluidized-bed Treatment of Water and Wastewater," Harwood., 1981.

4. M. Henze, "Anaerobic Treatment of Wastewater in Fixed-film Reactors," Pergamon, 1982.

5. G. F. Andrews, C. Tien, AIChE Journal, <u>27</u>, 396, 1981.

6. G. F. Andrews, J. Przezdziezcki, Biotech. Bioeng., in press.

7. G. F. Andrews, Biotechnology Progress, <u>2</u> 16, 1986.

8. G. F. Andrews, J. P. Fonta, "A Novel Adsorbing Bioreactor", 8th Symposium on Biotechnology for Fuels and Chemicals, Gatlinburg, Tenn. May, 1986.

9. G. F. Andrews, C. Tien, AIChE Journal, _25,_ 720, 1979.

10. Smith E. L., Greenshields R. N., Chem. Eng. _281,_ 28, 1974.

LARGE SCALE CULTURE OF HYBRIDOMA AND MAMMALIAN CELLS IN FLUIDIZED BED BIOREACTORS

Robert C. Dean, Jr.
Sabhash B. Karkare
Nitya G. Ray
Peter W. Runstadler, Jr.
Verax Corporation,
Lebanon, NH 03766

and

K. Venkatasubramanian
Department of Chemical and Biochemical Engineering
Rutgers University, Piscataway, NJ 08854

INTRODUCTION

Hybridoma cultures are being used routinely now for the in vitro production of monoclonal antibodies (MAb's). Mammalian cells are increasingly becoming the host cells of choice for the large scale production of an array of genetically-engineered therapeutic proteins such as tissue plasminogen activator (TPA). The manufacturing costs of those proteins is still very high. The application of novel bioreactor technologies is likely to reduce this cost significantly. In this paper, we address the design considerations and the operationl strategies of a new bioreactor system for the large scale, continuous production of these proteins.

CONTINUOUS CULTURE

The traditional hybridoma suspension batch culture commences at a cell density of about 1×10^5 cells/mL. Harvest titers can vary greatly; after 7-10 days, they are generally between 10-50 μg, of MAb/mL. The resulting volume productivity is about 0.4 mg of MAb/L of reactor volume/hr. [1]. This value does not count downtime for cleaning and sterilizing which can reduce volume productivity further by 10-20%. Batch reactors up to 1000 liters are now being used [1]. The transient nature of batch culture makes process optimization and control difficult, although the short cycle time renders aseptic practice less stringent compared to the requirements of continuous culture.

Long-term, fed-batch, free-cell cultures with periodic harvest extraction are practiced, some at a large scale (e.g., 8000 L reactor volume [2]). These systems have the unfortunate transient characteristics of batch systems and the aseptic-technology demands of continuous systems. Their performance lies between those of batch cultures and chemostats.

Chemostats operate with freely-suspended cells and have continuous nutrient feed and product harvest. We have operated chemostats, making MAb's, for over 150 days without contamination and producing up to 5 mg/L/hr,(space-time productivity). Cell concentration typically reach 2-4 x 10^6 cells/mL with viabilities of 60-90% depending upon the operating dilution rate. A typical optimum dilution for chemostats is 0.02 hr^{-1} giving about 3 x 10^6 cells/mL, 70% viability and a MAb harvest titer of about 240 mg/L. For steady state operation, the dilution rate should, of course, be well below the washout condition.

For a typical murine hybridoma, minimum cycle time is about 14 hours and washout occurs at a dilution rate of 0.049 hr^{-1}. At a dilution rate of 0.01 hr^{-1}, the cell cycle time is extended to 70 hours. With many of the hybridomas we have tested, there is an inverse relationship between cell growth and cell-specific MAb output rate. Figure 1 illustrates this effect for an IgG-producing

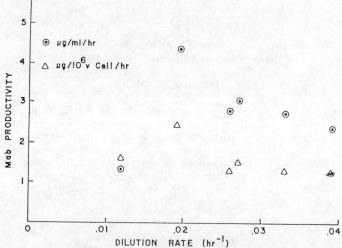

Figure 1. Inverse Relationship Between Cell-Specific MAb Output and Cell Growth Rate

hybridoma. Cell-specific productivity (μg, of Mab/10^6 viable cells/hr) more than doubles as the dilution rate was lowered, while viable cell density increased causing the reactor volume productivity to increase by more than two fold.

In an attempt to increase reactor space-time productivity by significantly increasing cell colony density, various schemes have been investigated to prevent cell loss from the chemostat. Himmelfarb, et al [3] and Feder and Tolbert, et al [4,5] have used rotating screen filters within the chemostat to prevent cell loss in the harvest stream. These investigators report cell densities of up to 2×10^7 cells/mL, but the reactor volume productivity did not increase in proportion, relative to the chemostat. Based upon the cells densities expected in chemostats, as discussed above, about 10 times more space-time productivity would be expected. But, in fact, reported data [5] (Table 1) was 2.6

TABLE 1
COMPARISON OF MAb PRODUCTION[13] PROCESSES

System	Productivity[7] (mg/LR[1]/hr)	Yield (g/LM[2])	Reactor Harvest Concentration (g/LH[5])	References
A) Mouse[11]	9×10^{-2}	na	$10^{(3,6)}$	Fanger [10]
B) Batch/Free-Cell	2	0.01-0.05	0.01-0.05	Phillips [11]
C) Chemostat	5	0.1-0.3	0.1-0.3	Verax Data, VX-2
D) Chemostat with Cell Return	2-3	-	-	Tolbert, et al [5]
E) Microtubule	10-15	-	-	Hosimer, et al [7]
F) Batch/ Encapsulated HY[4]	$0.79^{(10)}$	0.5	2.5	Littlefield [12]
G) Verax CF-IMMO[9]	25	0.1	0.1	Verax Data

1) LR = Reactor Volume (liters)
2) LM = Medium Volume Consumed (liters)
3) Ascites
4) Microshell-Encapsulated hybridoma
5) LH = Harvest Liquor Volume (liters)
6) Typical; ranges from 1 to 20
7) Gross Output; Not Accounting for Purification Loss (~30%)
8) Private Communication
9) From Verax CYTOCELL[TM] Bioreactors, Long-Term Average for Hybridomas VX-7 and VX-12
10) Anon; London Times, 26 July 1985 [9]
11) 25 µg/hr/mouse = Average for Colony
12) Confirmed by Birch, et al (Celltech Ltd), Trends in Biotechnology, (Elsevier), July 1985
13) All from Murine Hybridomas

mg/L/hr of MAb. Other investigators have used tangential flow microporous or ultrafilters in order to retain the cells. However, there is no report that these systems have operated for long periods without fouling.

The microtuble tube-in-shell type of reactor (e.g., the Amicon system, reported by Weisman, et al [6], and those of Bioresponse, Endotronics, Monsanto, A/G Technology, Microgon, etc.) is being widely used in laboratories in order to produce limited quantities of MAb's and mammalian cell proteins (MCP's). Data from independent researchers e.g., Hosimer [7] suggest that this type of system has a reactor productivity of about 10-15 mg MAb/L of reactor shell volume/hour, although manufacturers are claiming 2-3 times this level. These systems are easy to use, but eventually the cells overcrowd the extra-capillary space (shell side volume) producing a non-uniform cellular environment with significant gradients and restricted diffusion of nutrient and waste products. Amicon reports a reactor life of 6 weeks [6] although Bioresponse claims that operating times of over 12 months are achieved by washing excess cells from the extra-capillary space. Endotronics uses a shell side reservoir system to perfuse nutrients from the

128

capillaries through the extra-capillary space to the reservoir and then reserves the process in an attempt to achieve better distribution of nutrients in the shell side volume.

IMMOBILIZING WITH BEADS

Several systems have been developed with immobilized cells on or in microspheres. The "microcarrier" type spheres (e.g., Pharmacia's Cytodex microcarriers) are useful only with attachment-dependent cells which adhere to the surface of the microcarrier. There, they are exposed to collision damage in any fluidized or stirred bed type of reactor. In a packed bed reactor, the cells tend to fill the intersphere space and plug the bed.

Lim and Moss [8] developed a microsphere encapsulation technique used by Damon Biotech where the cells are inside a nutrient-permeable polylysine shell through which the cells and the product cannot escape. In 10-20 days, the cells crowd the capsule volume, either breaking through the shell or dying (maybe from restriction of nutrient and product diffusion). Reported reactor space-time productivity for MAb's is 0.8 mg/L/hr of reactor volume [9]. The encapsulation process also appears to be expensive. The high MAb harvest titers seen by Damon can be achieved from any of the other processes described by adding one step of ultrafiltration that can concentrate 10 to 100 times, according to our own production experience.

CELL IMMOBILIZATION IN MICROBEADS

We have developed a process for culturing cells that involves a continuous-operating fluidized bed of porous, sponge-like microbeads. The cells are colonized inside the beads, which are macro-porous through their surface so that excess cells and the products by cells, may escape. We have developed weighted microsponge beads (WS-IMMO^{TM} beads) made from a modified and strongly cross-linked natural polymer, and designated the continuous culture with these beads CF-IMMO^{TM} process.

The WS-IMMO^{TM} beads are given a high specific gravity by weighting them with heavy metals, especially selected to be noncytotoxic. At a specific gravity of 1.6, we can retain 90% of the bead volume open for cell colonization. Mammalian cells grow to very high densities in these beads. Typical cell densities are 1-2 x 10^8 and 2-3 x 10^8 cells/mL for lymphatic hybridomas and tissue cells, respectively. Highly productive cell colonies have been cultured in these beads for as long as we have run them (over 120 days). No limit on continuous-culture time, for a single charge of cells, has yet been observed. Because cell reproduction is not inherently required by the CF-IMMO^{TM} system, reproduction may be greatly reduced when working with cells that produce non-growth-associated products. This fact tends to extend significantly the genetic life of a highly productive culture. We have used both regulatory drugs and specific nutrient deprivation to control cell doubling time and have extended it up to 10 times the minimum cell cycle time (e.g., for hybridomas from about 14 hours to 140 hours). As mentioned above for chemostat cultures, in the case of some hybridomas, cell-specific productivity of MAb is inversely related to the cell growth rate; hence large increases in output are observed for these cells by slowing down cell reproduction.

THE FLUIDIZED BED BIOREACTOR

The inoculated WS-IMMO^{TM} microbeads are cultured as a thick slurry in a fluidized bed. The fluidized-bed approach is used in order to provide very high mass-transfer rates, especially for oxygen and carbon dioxide, in order to support CF-IMMO^{TM} 20-100 times conventional cell-colony densities. For example, we can supply oxygen at rates up to 40 mg-moles/L of reactor volume/hr. This is practical only with heavy (wet specific gravity > 2.0) WS-IMMO^{TM} beads that permit high recirculation rates, up to recycle dilution rates of 150 hr^{-1}. Without high mass-transfer and heat-transfer capabilities, the high cell colony densities inherent with CF-IMMO^{TM} could not be maintained adequately.

An equally important reason for adopting the fluidized bed is its ease of scale-up for mass production. CF-IMMO^{TM} systems have been scaled up 20 times, so far, with excellent predictability. The fluidized-bed bioreactor scales at constant bed depth. The preferred bed depth for production systems is around 2 meters.

THE RECYCLE BIOREACTOR

In order to suspend the thick slurry of beads in a fluidized bed, superficial velocities of 4mm/sec are used to 500 μm. specific gravity 1.4 beads. This fluidizing flux is extracted from the top of the bioreactor and returned, via a recycle loop, at the bottom. There is nothing in the bioreactor vessel except the fluidized bed and its stablizing distributor. Outside the reactor, the recycle liquor is conditioned in the recycle loop, comprised of a pump, a membrane gas exchanger to add 0^2 and CO^2, sensors for control pH, DO and temperature (using computer feedback) and a heater/cooler. The recycle circuit is shown schematically in Figure 2 and may be identified in the photograph of Figure 3.

Because 1 to 5% of the total cell population is present in the culture liquor - - while 95-99% are inside the WS-IMMO^{TM} beads _ _ the recycle loop components must be carefully designed so as not to damage the protein product. The temperature, shear stress, roughness, materials, etc. of internal surfaces must be controlled carefully in order to limit the cell death rate.

In the fluidized bed, the WS-IMMO^{TM} beads (and the cells immobilized in them) are treated very gently. We have observed little bead attrition after several thousands of hours of operation. The WS-IMMO^{TM} beads do not recycle; Figure 4 shows the very sharp disengagement at the top of the bed.

WS-IMMO^{TM}

The WS-IMMO^{TM} beads are made from a sponge-like polymer matrix having pores appropriately sized to mechanically entrap cells and allow them to colonize the beads.

Attachment-dependent cells, like fibroblasts, are also mechanically entrapped, but appear to attach readily to the polymer surfaces. In order to retain the cells adequately, but not to hold them so firmly that excess members of the cell colony cannot escape from the open pores in the bead surface, the correct matrix pore size and three-dimensional morphology must be provided. In the case of hybridomas, which are 10-15 microns in diameter and roughly spherical in shape, an appropriate matrix pore size is 20-40 microns. The channels in the sponge are interconnected so that the cells can enter easily into the entire sponge matrix during inoculation.

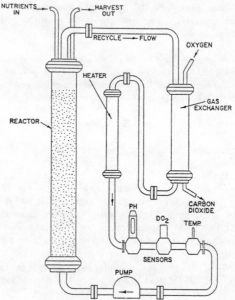

Figure 2.   Schematic of CF-IMMO^{TM} Plant

With hybridomas, the bed is inoculated as 2-5 X 10^5 cells/mL of reactor volume. The cells are slowly taken up by the beads and the free cell density drops markedly within a few days. Then, after 7 to 10 days of incubation, the cell colony inside the beads reaches densities of 5 X 10^7 to 1 X 10^8 cells/mL of matrix, while the cell density in the liquor surrounding the beads remains at 2-10 x 10^5 cells/mL.

The bead slurry density in the fluidized bed is typically 30-40% solids, which is expanded only slightly from the packed bed density of about 55% solids. However, even at this high slurry density, the beads are truly fluidized and move freely throughout the bed. These bead kinematics are essential in order to prevent cell growth between the beads, which would soon coagulate the bed.

131

Figure 3.　CF–IMMO™　Pilot Plant

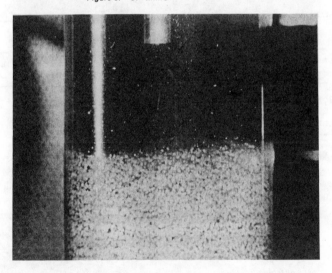

Figure 4:　　　Sharp Demarcation Between Top of
Fluidized Bed and Culture Liquor

In order to obtain the highest cell concentration in the bioreactor, one wishes to use the highest slurry solid fraction possible. For hybridomas, we have achieved bed-average cell colony densities of between 3 to 5 x 10^7 cells/mL of fluidized bed. The viability of a colony depends upon the culture-kinetic conditions and type of cell. Typically for hybridomas, we find it is 60-85% ; for tissue cells, 85-95%.

Observations of a number of CF-IMMO^{TM} cultures of hybridomas show that cell-specific productivity is not impaired by immobilization. Therefore, the CF-IMMO^{TM} hybridoma culture systems can achieve more than 20 times the reactor-volume space-time productivity of a free-cell suspension culture in either chemostat or batch mode. For typical hybridomas making murine MAb's, the space-time productivity is more than 25 mg/L/hr for the expanded bed volume (Table 2).

TABLE 2

OBSERVED PRODUCTIVITY IN CHEMOSTAT
AND CF-IMMO^{TM} HYBRIDOMA CULTURES

Product	Cultivation	Space-Time Productivity (mg/L/hr.)
IgG	Chemostat	1.3
IgG	CF-IMMO^{TM} (800 mL)	18.2
IgG	CF-IMMO^{TM} (10 L)	29.0

Cell Line	Cultivation	Cell Concentration (cells/mL/beads)
Mouse Hybridoma	CF-IMMO^{TM}	1.0×10^8
Mammalian Cell	CF-IMMO^{TM}	3.0×10^8

The morphology of the polymer sponge matrix can be varied using a proprietary matrix manufacturing process. The range of geometry is illustrated in figures 5a and 5b. Figure 5a shows a "wire-like" three-dimensional network while Figure 5b illustrates a "leafy" morphology.

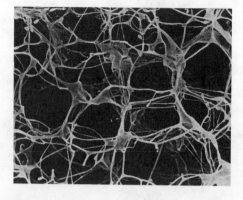

Figure 5(a). Wire-Like Matrix Structure

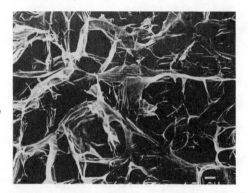

Figure 5(b). Leafy-Like Matrix Structure

Mammalian cells and hybridomas seem to prefer the "leafy" structure and achieve significantly higher cell colony densities in it than in the "wire-like" structure.

The basic manufacturing process involves casting the microbeads from a gel. The number-average diameter is about 500 microns. With a 40% slurry density, this gives 6 x 10^6 beads/L and a bead surface area of 4.8 m^2/L of expanded bed volume. Of course, the internal surface area of the beads is much higher; we estimate this to be about 300 m^2/L. Typically, 2% of the sponge matrix volume is polymer. About 9% of the volume is occupied by small metal weighting particles at specific gravity of 1.6. With 11% of the volume occupied by these materials, 89% of the WS-IMMO^{TM} microbeads is available for cell colonization at a wet specific gravity = 1.6. We have developed production techniques for making beads with a very narrow variation in diameter, over a range of diameters and pore sizes, and specific gravities from 1.05 to 2.5. While computer modeling of the culture process in the fluidized bed shows that approximately 200 microns is an optimum bead diameter, the optimization of the CF-IMMO^{TM} process is not a strong function of the bead diameter because, for mammalian cells, the oxygen consumption rate is relatively small and adequate oxygenation can be achieved for bead diameters up to 1000μm.

The matrix within each bead is strongly cross linked by a proprietary process. This cross linking makes the polymer matrix sufficiently stable to survive intact for many months in the CF-IMMO^{TM} bioreactor.

The WS-IMMO^{TM} beads are elutriated in order to establish their fluidization velocity within an acceptable range. Actually, the CF-IMMO^{TM} fluidized bed proves to be very tolerant, so that a precise fluidization velocity characteristic for the WS-IMMO^{TM} beads is not critical. Finally, the beads are sterilized. The micro beads are usually stored and shipped in a medium solution. Shelf life is at least six months.

CULTURE MEDIUM

Proprietary medium formulations have been developed which are especially tailored to the CF-IMMO^{TM} process and are designed to minimize the cost of downstream processing. For example, one successful formulation contains only 2μg/mL of protein. The composition is shown in Table 3 and typical, comparative

TABLE 3

COMPOSITION OF VERAX (DM-7-A)
LOW PROTEIN MEDIUM

Base:

    75% Dulbecco's Modified Eagles Medium

    25% Ham's F-12 Nutrient Mixture

Addition:

    10^{-9} Sodium Selenite
    10^{-4} M 2 - Aminoethanol
    10^{-2} M HEPES
    10^{-2} M Sodium Biocarbonate
    1 µg/mL Transferrin
    1 µg/mL Insulin
    46 µg/Liter α-Lipoic Acid
    19 µg/Liter Linoleic Acid
    10 mL/Liter R1642
    10 Units/Liter PFA-1

---

productivity results in Table 4. It is a completely defined medium, without fetal calf serum, bovine serum or serum albumin, so that the protein load on the downstream purification system is greatly reduced, as compared to conventional media. Our developments have shown that such well defined media are critical in order to make a detailed biochemical analysis of the culture liquor. There is a proprietary ingredient (R1642) in the medium formulation of Table 3. It is a defined chemical complex which assists is maintaining high levels of cell productivity.

In addition to the downstream processing and research advantages of our defined, essentially protein-free media, these formulations have greatly reduced the cost of nutrients for growing hybridomas, as shown in Table 5. Because of this large reduction, nutrient cost shrinks to a smaller percentage of total production cost.

MATERIALS OF CONSTRUCTION

In order to employ these new medium formulations in commercialproduction hardware, we have had to pay close attention to the material of construction. In general it can be said that, under certain conditions (the use of low protein medium for example), many common materials are toxic to hybridomas and mammalian cells. We have found, for example, that some Teflons, seal elastomers, bead-weighting materials, etc. are cytotoxic. After hundreds of biocompatibility tests (using a specifically-sensitive hybridoma cell line and a protein-free medium), we have been able to select and specify materials which demonstrate no cytotoxicity.

The most intense contact between the cells and materials is in the WS-IMMO™ beads.

TABLE 4 COMPARATIVE PRODUCTIVITY[5] USING SERUM CONTAINING[1]
SERUM-FREE, AND PROTEIN-FREE[3] MEDIA[4]

Medium	Day		
	1	2	3
DMEM[2]:			
Cell Number[6]	3.3×10^5	1.1×10^6	2.0×10^6
P^{+}[1]		0.48	1.43
DM-2A[3]:			
Cell Number	1.7×10^5	3.1×10^5	6.9×10^5
P^{+}		1.51	1.70
DM-7A[4]:			
Cell Number	2.0×10^5	3.2×10^5	6.3×10^5
P^{+}		1.35	1.79

1) P^{+} = µg of MAb/10^6 Viable Cells/hr.
2) DMEM: Dulbecco's Modified Eagles Medium
3) DM-2A: Verax Defined Medium/Serum-Free
4) DM-7A: Verax Defined Medium/Protein-Free
5) VX-7-0: Hybridoma Cell Line
6) Cell Number: Viable Cells/mL

TABLE 5

MEDIUM COST REDUCTION

Medium	Type	Ingredients ($/L)[5]	Labor & Overhead ($/L)	Total Cost ($/L)
Commercial:				
Liquid[1]	DMEM/10	41.55	3.50	45.05
Powdered[2]	DMEM/10	19.37	2.50	21.87
Serum-Free[3]	HB-102[2]	32.50	2.50	35.00
Verax:				
Serum-Free[4]	DM-2A[3]	1.75	2.00	3.75
Protein-Free[5]	DM-7A	2.25	1.75	4.00

(1) DMEM/10   Dulbecco's Modified Eagles Medium
                Plus 10% Fetal Calf Serum
(2) HB-102   Hanna Biological Defined Medium 102.
(3) DM-2A   Verax Defined Medium/Serum-Free
(4) DM-7A   Verax Defined Medium/Protein-Free
(5) $/L   of Liquid Medium

Figure 6 shows a mammalian cell actually growing on the surface of the weighting material used in the beads. The cell had extended filaments over the weighting material surface, which is a strong indication that it finds the surface to be benign. These small weighting particles are made by a unique proprietary process from a very carefully controlled alloy formulation. Many other weighting metals tested have produced dramatic toxic effects. In general, careful materials control is practiced for all of our CF-IMMO^{TM} systems.

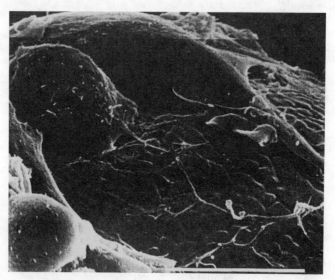

Figure 6.   Mammalian Cell Growing on the WS-IMMO^{TM}
            Bead Weighting Material

CULTURE KINETICS MEASUREMENT AND OPTIMIZATION

The CYTOCELLs^{TM} are 1 L (or smaller), complete CF-IMMO^{TM} reactors that replicate our pilot and production culture systems. We have found excellent correlation between results achieved in 500-mL, 1-L and 10-L CF-IMMO^{TM} reactors.

With the nutrient dilution rates (based on reactor volume) of the CF-IMMO^{TM} reactors being some 20 times or more higher than for a chemostat, the time to achieve fluid-dynamic, steady-state operating points can be reduced by a like factor. This provides a more practical time to achieve steady state for extensive investigation of the many parameters of the culture. Further, the precise control of the CYTOCELL^{TM} system is an essential feature. The pH is maintained to a precision of +5% of set point and temperature to +0.5% C by computer-operated, feedback control (using an IBM PC with a special interface, capable of multiplexing four CYTOCELL^{TM}. The recycle dilution rate, the nutrient dilution rate and the medium formulation can all be varied.

Additives may be fed separately to the reactor. In general, we do not vary the characteristics of the WS-IMMO^{TM} microbeads for mammalian cells, although they also could be tailored for a particular cell line or culture conditions. Assays are conducted by aseptic sampling of harvest liquor and beads. A special in - situ steam-sterilizable connector (VAC^{TM}) is employed for this purpose (and also for connecting and disconnecting medium supply, and harvest vessels and for removing sensors and pumps for maintenance and calibration, etc.)

The CF-IMMO^{TM} hardware system has been engineered for a high level of aseptic sophistication. We have operated CYTOCELL^{TM} reactors for well over 120 days without contamination.

For the culture kinetics work, extensive modeling of the culture is utilized. This work is now extending into modeling of the mammalian cell. This computer-based modeling is a powerful tool for the rapid discovery of optima in the complex function space. Figure 7 illustrates a typical improvement in a culture of hybridoma cells. hybridoma cells.

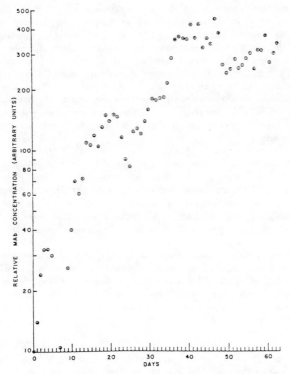

Figure 7.   Typical Improvement in a CF-IMMO^{TM} Culture of Hybridomas

When these cells were received they were producing a harvest concentration, in the chemostat mode, of less than 20 mg/L. After about six weeks of chemostat optimization work, the output titer increased by more than 20 times with a proportionally increased space-time productivity of the reactor. With the CYTOCELL^{TM} , such improvment can now be effected in a much shorter time.

FEATURES OF THE CF-IMMO^{TM} BIOREACTOR

The fluidized-bed reactor is a simple, constant-diameter cylinder. A proprietary distributor is employed at the bottom to stabilize the bed and to prevent matrix beads from flowing backwards out of the reactor, while permitting accidentally-recycled beads to re-enter the bed. The details of this design are important in order to prevent attachment-dependent cells from growing in and fouling the distributor passages.

The recycle pump located in the recycle loop is chosen so as not to damage recycling-free cells excessively. For the 10-L reactor systems culturing hybridomas, recycle dilution rates are typically 100 hr^{-1} or less, so the pump has a capacity of about 20 L/minute. The pump produces sufficient pressure rise to suspend the fluidized bed and to overcome the pressure drop of the recycle loop. The CF-IMMO^{TM} bioreactor is completely filled with liquid so there is no head space nor gas bubble in the system. The system has been designed specifically to avoid liquid-gas interfaces in order to prevent foaming, because this is known to damage certain delicate protein products. The reactor is pressurized to about one-half bar in order to force outflow, should any accidental leaks occur. The entire system is pressure-tested before and after in - situ steam sterilization (two bars absolute for one hour) and all leaks are corrected to the level where there is no observable pressure decrease within the aseptic envelope of the system over a period of 24 hours. Static O-ring seals are permitted, in order to top seal around sensors, and in sterile connection (TriClover or VCO) and at the end caps of glass reactor vessels.

All entry ports through the aseptic envelope of the system are protected by either a live-steam barrier (a "steam block") or by bacteriological absolute filters. All filters are steam sterilized and integrity tested before use. Gas-entry filters can remain on line for months. Liquid filters are changed periodically. Medium is sterilized through double, 0.1 micron, absolute filters. Medium is not quarantined; it is stored at 4°C. in the feed banks. The harvest tanks generally are sized to be changed once a day.

In - situ steam-sterilizable diaphragm metering pumps are used for the medium, for any special additives, and for base solution (used to control pH). The harvest liquor continuously flows out of the filled system at the same rate that the medium is pumped in. Several aseptic sampling ports are provided on the production culture module. Samples of the culture liquor or of WS-IMMO^{TM} beads from the reactor can be taken from these aseptic ports.

The tube and shell gas exchanger is of special design using silicone-rubber, tubular membranes. The culture liquor passes through the tubes at a high Reynolds number in order to minimize fouling by cell growth on the tubes. In the shell side of the gas exchanger, around the tubes, the partial pressures of oxygen and carbon dioxide are controlled. Thus, the dissolved oxygen and dissolved carbon dioxide in the bioreactor are maintained at precise set points. During in - situ situ steam sterilization of the gas exchanger, two bars absolute steam is placed both inside and outside the silicone rubber tubing, so it is subjected to no undue thermal/pressure stresses. The gas exchanger component is visible in Figure 3.

There are two recycle loops on the production reactor system; one is sterilized, medium filled and kept on standby. This insurance has been provided because conventional sensors of pH and DO prove to be somewhat unreliable and require cleaning and recalibration over the long periods of continuous operation. When sensor maintenance is necessary, the standby loop is brought into service and the idle loop is isolated by live steam barriers. When the instruments are serviced, the idle loop is resterilized, in place, with steam, pressure tested, filled with medium and put on standby. Because there are no probe penetrations into the biorector, all of the sensors and liquor-conditioning components are readily accessible in the recycle loop outside of the reactor. There is nothing in the reactor except beads, the cells, the culture liquor and the distributor.

The entire aseptic envelope of the reactor system -- i.e., the reactor, recycle loop and components, medium addition lines and medium vessel, base addition lines and base vessel, and harvest vessel -- is all in-place cleaned and steam sterilized prior to putting the system into operation.

CF-IMMO^{TM} systems must be supported by the usual infrastructure of pure water, pure steam, pure gas, waste disposal and downstream processing. Note that the CF-IMMO^{TM} culture system has the aseptic envelope completely closed and isolated from the environment, hence housekeeping strictures in the manufacturing plant are minimal. Although antibiotics are often used in the culture liquor, that is not necessary in order to maintain contamination-free operation. Indeed, our production systems have been validated by specifically introducing a bio-burden into the system prior to sterilization and then operating the system successfully without the use of antibiotics.

TYPICAL CF-IMMO^{TM} PERFORMANCE

When evaluating experimental data from tissue culture, the imprecision of state-of-the art assays must be duly considered. Assays for the measurement of performance are time consuming; constant vigilance must be maintained to perform them routinely and to assure accuracy. Figure 8 shows our estimate of attainable precision from ELISA assays, used for measuring product concentration, as a function of the number of assays on a single sample. Note that 10 separate assays on a single sample, with the present technique, yields a precision of only +/- 15%. The two largest sources of error are in the dilution process, where dilution rates may reach up to 1 x 10^4, and in the quality of standards. We are performing routinely hundreds of sample assays/week using automated dilution equipment and computeraided data reduction. Our production assay accuracy is not at the 10-15% level. For murine antibodies, a goat anti-mouse-immunoglobulin assay is employed. When an antigen to the antibody is available, it is possible to conduct an activity assay also. We also routinely idiotype our MAb output.

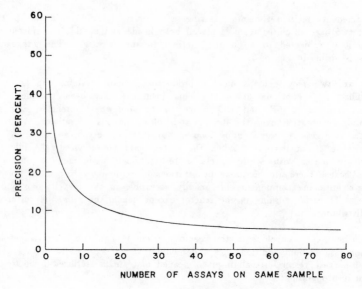

Figure 8. Best Estimate of Obtainable Precision
With ELISA Assay

Typical performance data for hybridomas producing MAb's is shown in Table 2 for both chemostat and CF-IMMO^{TM} modes (at several reactor sizes). Predictions of space-time scaled-up productivity are excellent. However, it is important to recognize that these CF-IMMO^{TM} cultures have not fully optimized, so there could be significant gains still available from perfecting the operating protocol for the CF-IMMO^{TM} systems. Table 1 compares the CF-IMMO^{TM} process to alternatives.

ECONOMICS

The cost of producing pure cell products on a large scale can be assessed only after considerable experience and analysis. The entire production system must be optimized and assessed together. For example, our protein-free medium greatly reduces the cost and increases the yield of downstream product processing, as well as reducing the cost of medium by as much as 10 times. However, if the medium used demands a high level of control of materials of construction, the cost of hardware may increase. The advantages of high product quality, automation and control simplification offered by continuous, steady production tend to be offset by the requirements of rigid aseptic management.

We have constructed a global computerized model of a CF-IMMO^{TM}, MAb/MCP production plant involving 19 multiple-variable modules. This is being used to optimize plant design for specific products. With such a computer model, bottlenecks become obvious and means to remove them, in order to lower production cost, can be tested quickly. Without computer-modeling, the optimization of the large number of variables will be at best intuitive. Because of the wide range of products of interest in the high value protein market, the plant configuration often changes significantly for each specific product manufacturing campaign. Experience-based human intuition, traditionally employed to optimize complex processes, tend to be ineffective. Only after many production campaigns, requiring several years of commercial output, will we be able to assess with certainty the economic advantages of the CF-IMMO^{TM} process and to compare it without bias to alternative processes.

CONCLUSION

We have developed a unique production process for the mass culture of mammalian cells. Commercial-production results with this process are, to date, very encouraging, but are yet based upon limited pragmatic experience in large-scale industrial production. Time will tell whether it is the preferred approach. Its advantages and its strengths are being enhanced by continuous research and development. With our CF-IMMO^{TM} process, mammalian cells are now being cultured in a systematic, technologically-definitive fashion. By applying engineering discipline to the art of tissue culture, it has been possible to measure and repeat data under precise control of the great number of variables, to predict scale-up requirements for industrial scale production, and to utilize components and systems engineered for the industrial environment. We expect that this technology will generate major reductions in the cost of pharmaceutical proteins.

ACKNOWLEDGEMENT

The authors are greatful to Mrs. Susan Koontz for her assistance in the preparation of this article.

REFERENCES

1. Thompson, P. W., L. A. Wood, J. R. Birch, K. Lambert, and R. Boraston. Antibody Production in Airlift Fermenters; American Institute of Chemical Engineers Annual Meeting, Chicago, Illinois, November 10-14, 1985.

2. Phillips, A. W., G. D. Ball, K. H. Fautes, N. B. Finter and M. D. Johnston. Experience in the Cultivation of Mammalian Cells on the 8000 Liter Scale. Large-Scale Mammalian Cell Culture Edited by J. Feder and W. R. Tolbert; Academic Press, Inc., 1985, pp. 87-96.

3. Himmelfarb, P., P. S. Thayer, and H. E. Martin. Spin Filter Culture: The Propagation of Mammalian Cells in Suspension; Science 164; 55-557, 1969.

4. Feder, J. and W. R. Tobert. The Large-Scale Cultivation of Mammalian Cells; Scientific American, 248: No. 1:36-43, 1983.

5. Tolbert, W. R., Jr. C. Lewis, P. J. White, Feder, J. Perfusion Culture Systems for Production of Mammalian Cell Biomolecules: Large Scale Mammalian Cell Culture. Edited by J. Feder and W. R. Tolbert; Academic Press, Inc., pp. 97-123, 1985.

142

6. Weisman, M. C., B. Creswick, P. Calabresi, J. Hopkinson, R. S. Tutuyian. Cellular and Biochemical Aspects of Human Tumor Cell Growth and Function in Hollow Fibre Culture; Large-scale mammalian Cell Culture Edited by J. Feder, and W. R. Tolbert, Academic Press, Inc., pp. 125-126, 1985.

7. Hosimer, P. Private Communication. Eastman Kodak Company.

8. Lim, F. and R. D. Moss. Microencapsulation of Living Cells and Tissues. J. Pharm. Sci., 70, No. 4:351-354, April 1981.

9. Anon; London Times, 26 July 1985.

10. Fanger, M. W. (Dartmouth Medical School); Private Communication: 1983.

11. Phillips, S. M. University of Pennsylvania, School of Medicine, Philadelphia, Pennsylvania; Private Communication. 1983.

12. Littlefield, S. G., K. J. Gilligan and A. P. Jarvis. Growth and Monoclonal Antibody Production from Rat X Mouse Hybridomas: A Comparison of Microcapsule Culture with Conventional Suspension Culture. Damon Biotech Inc., Abstracts from the Third Annual Congress for Hybridoma Research, February 27, 1984, San Diego, California, Hybridoma, 3: No. 1:75, 1984.

EFFECT OF WATER MISCIBLE SOLVENTS ON BIOTRANSFORMATIONS
WITH IMMOBILIZED CELLS - AN OVERVIEW

A. Freeman, Department of Biotechnology
Faculty of Life Sciences, Tel-Aviv
University, Tel-Aviv 69978, Israel

ABSTRACT

One of the intrinsic problems often encountered with biotransformations
carried out by immobilized cells is the poor solubility of the substrate,
cosubstrate and product (e.g. steroids) in the aqueous medium. The low
solubility imposes severe diffusional limitations, resulting in low conversion
rates. Vigorous mixing is thus becoming necessary. The resulting shear forces
often lead to disintegration of the fragile beads containing the cells.

One of the obvious ways to allow for increased solubility of such substrates
is the replacement of a fraction of the aqueous medium by water miscible solvents.
Criteria for solvent selection for such applications and better understanding of
the effects of the solvent added on the catalytic activity, substrate solubility
and operational stability of the system involved might thus allow for better use
of such solvents with minimal side effects.

In this presentation criteria for solvent selection is described, as
elucidated from existing data on solvophobic effects and effects of solvents on
enzyme stabilization and destabilization. This criteria was tested on glucose
oxidase and esterase as free enzyme models, on Δ^1 dehydrogenation by *Arthrobacter
simplex*, on Δ^1 reduction by *Mycobacterium sp.* and on the glycolytic activity of
yeasts. Our results offer guide lines for solvent selection, as well as better
understanding of the mode of action of the added solvent on the kinetic parameters.

Better insight into the mechanism of stabilization of entrapped cells in the
presence of added solvents, often exhibited by the surrounding matrix, **was allowed
by** studies carried out on yeasts immobilized in crosslinked polyacrylamide-
hydrazide. (Key words: solvent effect; immobilized cells; steroid biotransformation).

143

Introduction

A distinctive group of bioconversions carried out by microbial enzymes is steroid modification (1). Many activities of oxidoreductases operating from a whole cell have been demonstrated, investigated and, in some cases, found useful on an industrial scale. No wonder then that conversions of such processes into immobilized cell based systems have been attempted in recent years. As gel entrapment of microbial cells allows for confining large amounts of biocatalyst within a small volume of reactor, entrapment in hydrophylic matrices (e.g. alginate, polyacrylamide) was mostly employed. Due to the low solubility of many steroidal compounds (mostly within the range $1-10 \cdot 10^{-5}$M (2) the added substrate is suspended in the medium. As direct uptake of steroid substrate by the cells is impossible in the case of entrapped cells, only dissolved substrate molecules which reach the gel surface diffuse through the hydrophylic matrix, reach the cell and undergo the envisaged conversion. Diffusional limitations (in the bulk medium and inside the gel) may thus play a major role in determining the reaction rate. Three major approaches were attempted, all aiming towards improvement of mass transfer: replacement of most of the bulk aqueous medium with water immiscible solvents (3,4), replacement of a fraction of the aqueous medium with water miscible solvents (5) and introduction of hydrophobic elements into the gel matrix (6). The introduction of water miscible solvents may significantly affect the system by both increasing the solubility of substrate and co-substrate and facilitating its migration inside the gel via swelling of gel matrix and creation of homogeneous medium inside and outside the gel. The penetration of solvent molecules into the gel leads to direct contact with the immobilized cell. Thus, effects of the co-solvent are expected on the kinetics and stability of the enzymes involved. Added cosolvents may, therefore, facilitate reaction rates due to the increased solubility but simultaneously may inhibit the reaction by its interaction with the enzyme(s) involved, or even inactivate the enzyme(s) either immediately or gradually. An important technical advantage expected from the use of cosolvents is that vigorous mixing is not required due to the increased solubility and creation of homogeneous medium. As vigorous mixing leads to disintegration of the fragile beads or gel fragments containing the cells, minimal or no mechanical mixing is essential for long term operation.

Several reports in literature describe the use of water miscible solvents as co-solvents employed to facilitate steroid transformations by immobilized cells (5, 7, 8, 9). Out of the very long list of water miscible solvents four solvents were mainly employed: ethanol (7), methanol (5, 8, 9), dimethylformamide (DMF) and dimethylsulfoxide (DMSO) (5). The selection of these solvents was carried out on an empirical basis. Sonomoto and co-workers (8, 9) tested the effect of a series of alcohols on Δ^1 dehydrogenation of hydrocortisone by acetone dried *Arthrobacter simplex* cells, immobilized in photo-crosslinkable resin or polyurethane. They concluded that methanol (5-10% (v/v)) is the solvent of choice (6, 8, 9).

In view of the potential inherent in the use of cosolvents, better insight into its mode of action on biotransformations with immobilized cells in general and set up of criteria for solvent selection in particular, seems to be required. In the following, these aspects will be discussed and tested on model systems.

Criteria for solvent selection

The introduction of a water miscible solvent into the reaction medium may affect the system in three major aspects: the solubility of substrate and product; the transport to and inside the gel; the kinetics and stability of the enzyme(s). As immobilized cell systems are expected to operate for long periods of time (in the range of weeks to months) the effect on the stability of the enzyme(s) involved turns out to be critical. Solvent selection should, therefore, be primarily based on its effects on enzyme stability. Effects on enzyme kinetics, mass transfer and solubility should be taken into account following this primary selection.

Guide line for solvent selection, on the basis of high retention of the enzymic activity, is available in literature for purified enzymes. Martinek and Berezin (10) recommended solvent selection on the basis of solvent effect on hydrophobic interactions within the enzyme molecule, which play major role in the denaturation process. Obviously, solvents which interfere with these interactions should be avoided; solvents which do not interfere, or even strengthen hydrophobic interactions, thus stabilizing the active enzyme conformation, are favoured. The

assumption that strengthening of hydrophobic interactions is a key parameter
in enzyme stabilization is supported by Back *et al* (11).

Data presented by Ray (12) for solvophobic interactions, expressed as
critical micelle concentration of a non ionic detergent in various solvents,
allowed for the division of water miscible solvents into two major groups - one
comprised of solvents such as water, glycerol, ethylene glycol, propylene glycol
and butylene glycol, all having two or more potential hydrogen bonding centers and
inducing micelle formation; the other is comprised of solvents such as dimethylformamide, methanol and ethanol which did not induce micelle formation on the
detergent tested. The data (Table 1) indicates interference of DMF, methanol and
ethanol with hydrophobic interactions. Moreover, it presented quantitative
estimation of the intensity of solvophobic interactions within the polyols group
of solvents.

Table 1: Solvophobic effects of some water miscible solvents (extracted from (12))

Solvent	Critical micelle concentration of "NPE$_9$" (nonionic detergent) (mol fraction)
water	$5.6 \cdot 10^{-5}$
glycerol	$8.7 \cdot 10^{-5}$
ethyleneglycol	$1.25 \cdot 10^{-2}$
1,3 butanediol	$4.0 \cdot 10^{-1}$
propyleneglycol	$5.0 \cdot 10^{-1}$
dimethylformamide	-
methanol	-
ethanol	-

The guideline and data presented above reflects an intrinsic problem inherent
in the use of cosolvents: increased solubility of hydrophobic substrate is desired
but with minimal or no effect on hydrophobic interactions within the enzyme(s)
involved. Thus, increase in the capability of an added solvent to retain hydrophobic materials solubilized in the aqueous medium might be accompanied by parallel
increase in its interference with the natural active conformation of the enzyme(s).
Hence, selection of cosolvent for a specific biotransformation reaction should be
made out of a group of mild solvents (e.g. diols) listed according to gradual
increase in its interference with enzyme stability.

In order to test this hypothesis the effect of diols and methanol, ethanol, DMF and DMSO (solvents previously employed as cosolvents for immobilized cell systems) was studied employing two purified enzyme models and two cell models.

Enzyme models

In order to test whether the relative effect of water miscible solvents on enzyme stability is in accord with the pettern reported by Ray (elucidated from solvophobic effects) two purified enzyme models were investigated (13).

As the first model glucose oxidase (E.C.1.1.3.4), catalysing the oxidation of glucose into gluconolactone, was chosen. Prior to stability measurements preliminary study of the effect of the solvents involved on the kinetics of the enzyme was carried out, aiming to define what solvent concentration may present in activity measurements without significant interference with the activity recorded.

Glucose oxidase is highly specific (14); solvent effects are therefore expected in this case to reflect mainly conformational changes imposed on the enzyme by the present cosolvent. The effect of a series of cosolvents (concentration range 0-8 M ($\sim$30% (v/v)) on Vmax and Km values is shown in Fig. 1 (A & B).

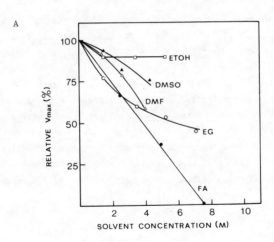

B

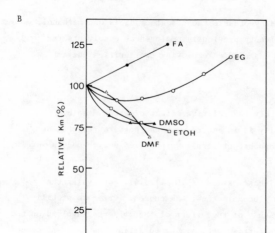

FIGURE 1. Effect of cosolvents on Km (B) and Vmax (A) of glucose oxidase
(from 13).

Kinetic measurements were carried out in presence of the indicated cosolvent
concentrations at 25°C, employing glucose as the substrate. FA: formamide;
EG: ethylene glycol; DMSO: dimethylsulfoxide;ETOH: ethanol; DMF: dimethylformamide.

The data of Fig. 1-A show that a decrease of 20-50% in the value of Vmax
was recorded for the solvents tested. The data of Fig. 1-B showed mild effects
on Km, depending on the nature of the solvent employed. These effects were found
to be fully reversible following the removal of the cosolvent: kinetic parameters
measured for glucose oxidase, physically entrapped in glyoxal crosslinked polyacrylamide-hydrazide (PAAH) gel (15), were effected by DMF and ethyleneglycol as found
for the free enzyme. The immobilized enzyme regained its initial Km and Vmax values
following washing of the cosolvent (Table 2).

Table 2: Reversibility of co-solvent effect on kinetics of PAAH entrapped
Glucose oxidase

	Conditions	Km(M)	Vmax (U/ml)
(A)	(1) Buffer	0.019	0.16
	(2) 50% (v/v) EG (24 hrs incubation)	0.023	0.04
	(3) Buffer (following washing)	0.018	0.16
(B)	(1) Buffer	0.019	0.086
	(2) 30% (v/v) DMF (24 hrs incubation)	0.0054	0.046
	(3) Buffer (following washing)	0.019	0.080

Our results indicate that in the presence of high cosolvent concentration
($\sim$5-50%) significant reduction in activity was recorded. Stability studies,
based on residual activity measurements were, therefore, appropriately designed
to ensure (by dilution) low content (less than 3%) of cosolvent in the reaction
mixture employed for activity measurements.

In order to measure the relative effect of cosolvents on the stability of
glucose oxidase, kinetic measurements of thermal inactivation of the enzyme were
carried out. To ensure fast removal of solvent prior to the residual activity
measurement, PAAH entrapped enzyme was employed. Plot of ln Vt/Vo vs time
(Vt: activity measured at time "t", Vo: initial activity (t = o)) at constant
temperature (55°C) gave a straight line, the slope of which is designated as
constant of inactivation (Kina) (16). Therefore, high values of Kina reflects
faster inactivation rates. Table 3 presents the effect of the same concentration
(3.5 M, $\sim$ 20%) of a series of cosolvents (two diols, DMSO, DMF and two alcohols)
on Kina of glucose oxidase at 55°C.

Table 3: <u>Effect of co-solvent on thermal inactivation of PAAH entrapped Glucose oxidase</u>

Solvent (3.5 M)	K_{ina} (h^{-1}) ($55^{o}C$)
none (0.2 M Pi, pH=6)	0.088
ethyleneglycol	0.053
propyleneglycol	0.093
dimethylsulfoxide	0.113
dimethylformamide	0.27
isopropanol	1.86
ethanol	5.38

The data of table 3 show that the pattern elucidated from Ray's study on solvophobic interactions was found for cosolvent effect on glucose oxidase thermal inactivation. The effect of solvent concentration on Kina for four representative solvents is shown in Fig. 2.

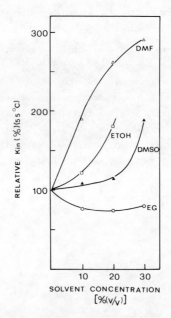

FIGURE 2. <u>Effect of cosolvent concentration on kinetics of thermal inactivation of glucose oxidase (13)</u>

151

The data of Fig. 2 show that ethylene glycol, a distinctive representative
of the diols group of solvents, had mild stabilizing effect throughout the
concentration range employed. A pattern of ethyleneglycol > DMSO > ethanol > DMF
was revealed. In the case of ethyleneglycol and DMSO concentrations of up to
20% had no significant effect on the stability.

As a second enzyme model, the enzyme pig-liver esterase (E.C.3.1.1.1)
was selected. This enzyme, which differs greatly from glucose oxidase in structure
and function, is effected in a very different way as far as its kinetic parameters
are considered (employing p-nitrophenylacetate as substrate) (13). Higher Km
values (up to x10) were recorded for all solvents tested. Significant increase
in Vmax (up to x2) was recorded for all diols employed, while DMF, isopropanol and
DMSO inhibited the reaction when concentrations higher than 10% were employed.
These effects were reversible, in accord with the results described above for
glucose oxidase. In spite of the different effects of cosolvents on the kinetics
of pig-liver esterase, the effect on its thermal inactivation was similar to that
recorded for glucose oxidase. Table 4 presents Kina values recorded for the same
group of solvents at the same concentration and temperature (55°C).

Table 4: Effect of cosolvent on thermal inactivation of pig liver esterase

Solvent (3.5 M)	Kina (h^{-1}) (55°C)
none (0.05M tris, pH=8)	0.268
ethyleneglycol	0.441
dimethylsulfoxide	0.460
propyleneglycol	0.779
ethanol	5.631
dimethylformamide	n.c.*
isopropanol	n.c.*

* n.c.: not calculated - total loss of activity within 5 min.

The data of Table 4 resemble that of Table 3, with the same pattern. The effect of cosolvent concentration on Kina of pig liver esterase is presented in Fig. 3.

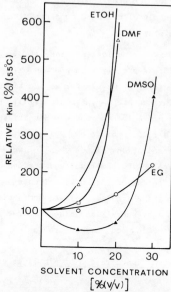

FIGURE 3: Effect of cosolvent concentration on kinetics of thermal inactivation of pig liver esterase (13)

The data of Fig. 3 show a similar pattern to that of Fig. 2. Although the intensity of the effect imposed by the cosolvent is greater, ethyleneglycol still reflects the mildest effect on stability throughout the wide concentration range tested. Again, ethyleneglycol and DMSO, when employed at concentrations up to ∿25% have very little effect on enzyme stability. Ethanol and DMF already had hazardous effects at ∿15%.

The data presented above for the two enzyme models confirm the hypothesis elucidated from the results on solvophobic interactions: diols are preferable as cosolvents, from the point of view of enzyme stability. DMSO comes next, almost equivalent, but other solvents previously employed as cosolvents for steroid

conversions by immobilized cells - (DMF, alcohols) seem to have hazardous
effects on enzyme stability.

   Although the above conclusion is based on data accumulated from only
two enzyme models, it seems to represent a general rule. In the following, the
effect of the same series of solvents on intracellular oxidoreductases, performing
bioconversions of steroids from a whole cell, will be described and discussed in
view of the results described above for purified enzymes.

Δ^1-Reduction of steroids by *Mycobacterium sp.* cells (18)

Δ^1-Reduction of 1,4-androstanediene-3,17-dione (ADD) to 4-androstane-3,17-dione
(AD) by *Mycobacterium sp.* NRRL 3805 was selected as the first cell model. The
reductase involved was reported to operate in the soluble cell fraction and NADPH
dependent (17). *Mycobacterium sp.* NRRL 3805 cells were grown under optimal
conditions, harvested and immediately frozen in liquid nitrogen. Prior to use,
cells were carefully defrosted in the cold. The effect of cosolvents added
(20% (v/v)) on reaction kinetics and residual viability of the cells - in
suspension - is presented in Table 5.

   The data of Table 5 clearly confirm the working hypothesis: the cosolvents
tested were divided into two groups - one comprised of diols and DMSO, which did
not impair cell viability but to some extent inhibited the activity, and another
comprised of DMF, alcohols, acetone, dioxane, acetonitrile and formamide, which
killed the cells and completely inhibited the reductase activity. It is pertinent
to note that DMF and methanol seem to represent a "transition" group of solvents,
small number of cells survived the 5 hrs incubation, but no activity could be
detected. The data of table 5 thus confirm that the same trend and pattern of
effect (diols > DMSO > DMF and alcohols) elucidated from solvophobic effects and
stability measurements with purified enzymes, appears in the case of the intracellular
enzyme operating from a living whole cell.

   The results presented above, as well as the data recorded for glucose oxidase,
raise the dilema involved in the use of cosolvents for facilitating bioconversions
of water insoluble substrates: desirable increased solubility, achieved with the
higher contents of cosolvent, may be accompanied by inhibition of the enzymic activity.
An optimum of cosolvent concentration is therefore expected at the point where
the increase in specific activity allowed by the increased substrate solubility

154

Table 5: Effect of co-solvents on ADD reduction by freshly
thawed *Mycobacterium* sp. NRRL 3805 cells (18).

Solvent (20% (v/v))	Relative activity[a] (%)	Viability[b]
none (50 mMPi; pH = 7, 0.1mMADD)	100	+
ethyleneglycol	81	+
glycerol	63	+
diethyleneglycol	54	+
triethyleneglycol	48	+
propyleneglycol	47	+
dimethylsulfoxide	43	+
dimethylformamide	0	v.l
methanol	0	v.l
ethanol	0	−
isopropanol	0	−
formamide	0	−
acetonitrile	0	−
dioxane	0	−
acetone	0	−

a) Reaction mixture included 8.5 mg (dcw) of cells in 20 ml of 0.1 mM ADD
in 50 mM phosphate, pH 7, shaken at 30°C (130 spm) for up to 5 hrs. Under
these conditions the substrate is water soluble. Kinetics was followed by
HPLC analysis.

b) One tenth of ml of reaction mixture was removed after the 5 hrs incubation
and transferred to MY agar plate (17). (+) indicates growth; (−) indicates
no growth; v.l. indicate very small number of colonies developed.

is not cancelled by the inhibitory effect of the cosolvent. In order to evaluate the maximal Δ^1 reductase specific activity possible by the addition of cosolvent to the system, the solubility of ADD in ethyleneglycol-water mixtures (room temperature) was determined (Table 6)

Table 6: Solubility of ADD in EG-water medium

EG content (v/v)	Solubility	
	g/l	mM
99%	26.9	–
30%	1.44	5.1
20%	0.96	3.4
10%	0.48	1.7
0	0.03	0.1

The data of Table 6 presents an increase of up to x 50 in soluble ADD by the introduction of up to 30% ethyleneglycol. The increase in solubility was linear with the cosolvent content. In order to optimize the cosolvent concentration allowing for maximal specific activity, the activity of same amount of cells, in presence of various ethyleneglycol concentrations, was measured (Table 7).

Table 7: Optimization of specific activity (μmol/g/h) of freshly thawed *Mycobacterium sp.* NRRL 3805 cells

	Ethyleneglycol content (% v/v)			
ADD(mM)	0	10	20	30
1.7	14.6[*]	31.0	28.3	26.1
3.4	28.7[*]	–	38.0	37.3
5.1	12.0[**]	–	–	32.8

[*] ADD insol. (fine suspension)

[**] ADD insol. (heavy precipitate)

The data of Table 7 confirms the inhibitory effect of ethylene glycol: employing 1.7 mM ADD (soluble at 10% EG) increase in cosolvent concentration led to decrease in specific activity. However, maximum activity was recorded with 3.4 mM ADD in presence of 20% EG, indicating the superiority of the cosolvent containing optimal combination over substrate suspensions. Moreover, higher relative increase in activity may be expected in the case of gel entrapped cells. The activity recorded for ADD suspensions (no EG present) revealed increase in activity by increasing the ADD input from 1.7 mM "concentration" to 3.4 mM. In both cases fine substrate suspension was formed. Sharp decrease in activity was recorded when ADD input was raised to 5.1 mM (heavy precipitate). These results indicate direct uptake of substrate by cells adsorbed to suspended ADD particles. This mode of substrate uptake is not possible in the case of entrapped cells, therefore the response of immobilized cells to added cosolvent may be expected to be even larger than that found for the freely suspended cells.

Δ^1 Dehydrogenation of steroids by *Arthrobacter simplex* cells (19)

Δ^1-Dehydrogenation of hydrocortisone to prednisolone by *Arthrobacter simplex* cells was employed as a second cell model. The dehydrogenation reaction has absolute requirement for oxygen as terminal acceptor of the hydrogen atoms. The inducible, membrane bound Δ^1-dehydrogenase is believed to have FAD as cofactor and is highly unstable in purified form. *Arthrobacter simplex* cells were grown and induced under optimal conditions, harvested and frozen in liquid nitrogen. Prior to use cells were defrosted carefully in the cold. The effect of cosolvents on the Δ^1-dehydrogenation activity of freshly thawed cells - in suspension - is presented in Table 8.

The data of Table 8 confirms once more the expected trend and effect of added cosolvents. Here again all cosolvents inhibited the dehydrogenation reaction to some extent, the order being the same as found for Δ^1-reductase activity of *Mycobacterium sp*, and as elucidated from the thermal inactivation studies with the purified enzymes. These findings indicate, therefore, that the criterion offered for solvent selection is of general nature. Once more the diols, followed by DMSO, are the group of solvents from which final selection should be made. Alcohols and DMF completely inhibited the dehydrogenation reaction. As ethyleneglycol was the first diol to allow for

157

Table 8: Δ^1-Dehydrogenation of hydrocortisone (1g/1;2.8 mM) by freshly thawed *Arthrobacter simplex* cells in presence of 20% (v/v) cosolvents [a] (19)

Solvent	mg substrate converted[b]	relative activity (%)	substrate
none (0.05 M tris, pH=7.8)	13.2	100	insoluble
glycerol	10.3	78	insoluble
ethyleneglycol	6.8	52	soluble
1,2 propanediol	5.3	40	soluble
1,3 butanediol	4.7	36	soluble
dimethylsulfoxide	3.0	23	soluble
dimethylformamide	0	0	soluble
isopropylalcohol	0	0	soluble
ethanol	0	0	soluble

a   Thirty mg of hydrocortisone were reacted with 10 mg (dcw) of freshly thawed cells, in 0.05 M tris, pH=7.8, shaken at 30° (150 rpm).
b   mg substrate converted per flask within 60 mins.

substrate solubilization under the conditions employed, with minimal inhibition, it was selected again as the cosolvent of choice. The capability of added ethyleneglycol to allow for increased solubility of hydrocortisone in EG-water mixture is presented in Table 9.

Table 9: Solubility of Hydrocortisone in EG-water medium

EG content (v/v)	Solubility g/l	mM
99%	10.0	-
30%	2.7	7.46
20%	1.8	4.97
10%	0.9	2.49
0	0.3	0.83

The data of Table 9 show an increase of up to x 9 in hydrocortisone solubility due to the introduction of the cosolvent (up to 30%). As in the case of ADD, the increase in solubility is linear with the amount of EG added.

The addition of EG had an inhibitory effect on the kinetics of the dehydrogenation reaction carried out by freshly thawed, suspended *Arthrobacter simplex* cells. In the case of 0.3 g/l hydrocortisone (maximal solubility in water) as well as in the case of 0.9 g/l (maximal solubility in 10% EG), addition of ethyleneglycol led to reaction rates lower than those recorded for the aqueous system (67-70% for 10% EG; 58-60% for 20% EG and 45-51% for 30% EG). A possible explanation for this finding is the effect of cosolvent on oxygen solubility and transfer. Recent studies on the effect of sugar concentrations in fermentation broth, on oxygen solubility, revealed decrease in oxygen solubility (20). In order to compensate for this effect an artificial electron acceptor - phenazine-methosulphate (PMS) - was introduced to the system (9). Employing optimal concentration of PMS (0.4 mM) a seven fold increase in the activity was achieved in presence of 20% EG. The data of Table 10 presents the effect of added PMS; conversion rates in presence of cosolvent were thus higher than those recorded with suspended, insoluble substrate.

Table 10: Effect of added 0.4 mM PMS on Δ^1-dehydrogenation of Hydrocortisone (1,8 g/l; 5 mM) by freshly thawed *Arthrobacter simplex* cells[*]

EG content (v/v)	PMS	Substrate	Relative Activity
0	0	insoluble	100% [**]
0	0.4 mM	insoluble	178%
20%	0	soluble	33%
20%	0.4 mM	soluble	233%

*   For conditions see legend to Table 8
** 100% = 2.1 mg/mg(dcw)/h

The second whole cell model thus confirm the working hypothesis. Although only two enzyme models and two whole cell models were demonstrated, the data seem to present a general criteria for cosolvent selection and better understanding of its effect on the kinetics and stability of the enzymes involved - either purified or operated from the whole cell.

Cell immobilization in polyacrylamide-hydrazide (PAAH) gels: effect of
surrounding matrix on the tolerance to cosolvents.

A mild method for immobilization of whole cells by gel entrapment
in crosslinked synthetic gel was developed in our lab in recent years
(for review see (21)). The method is based on suspending the cells in
aqueous solution of PAAH, followed by crosslinking with glyoxal (scheme 1).

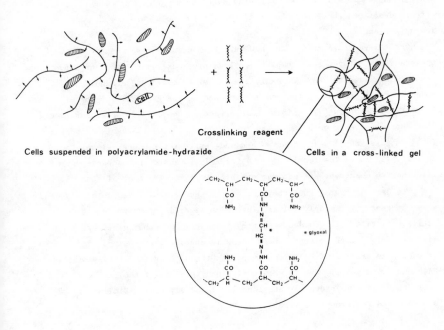

Crosslinking reagent

Cells suspended in polyacrylamide-hydrazide

Cells in a cross-linked gel

Scheme 1. Gel entrapment of cells in crosslinked polyacrylamide-hydrazide.

This method was found to be useful for the immobilization of microbial cells, yeast, plant cells, liver microsomes and following some modification, enzymes too (15,20). This highly porous, stable gel system allowed for high (mostly   90%) retention of immobilized activity (20). No wonder then that immobilization of both cell models described above could be carried out without impairing the immobilized activity of both *Mycobacterium sp* and *Arthrobacter simplex* cells. Characterization of these PAAH entrapped cell systems, at the immobilized state, and possible interference of the surrounding gel on cosolvent effects described above for the freely suspended cells, are now under investigation.

One relevant aspect should, however, be mentioned here even though the study has not yet been completed. In a previous report from our lab, yeast immobilized in PAAH gel exhibited higher tolerance to ethanol concentrations present in the medium surrounding the gel (22). In order to test whether this gel exhibits this effect in the case of other cosolvents PAAH entrapped yeast were incubated for one hour (30°C) in cosolvent-buffer (0.3% citrate, pH 5) solutions, washed and tested for residual glycolytic activity (22,23). The results (Fig 4) clearly indicate that enhancement of tolerance to cosolvent effects was exhibited only in the case of ethanol.

The data of Fig. 4 show that out of the second less favourable cosolvent group ethanol was the only solvent for which significant shift in residual activity profile was recorded. As ethyleneglycol and other diols are significantly milder solvents three hours incubation was required to observe irreversible effects of these cosolvents on the glycolytic activity of the PAAH entrapped yeast. Fig. 6 show a distinctive shift of the residual activity profile recorded for ethyleneglycol and glycerol. Similar effect was recorded for propyleneglycol (23).

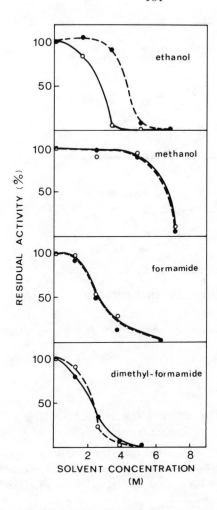

FIGURE 4. Test for protective effect of PAAH gel on immobilized
yeast: cosolvents from group II.

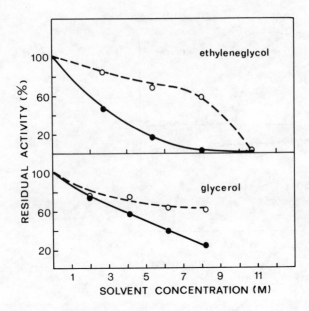

FIGURE 5. <u>Test for protective effect of PAAH gel on immobilized</u>
<u>yeast: cosolvents from group I.</u>

The data of Fig. 6 indicate a protective effect exhibited by the
PAAH gel. This finding raises the potential of employing high cosolvent
concentration in the bulk medium outside the gel while the entrapped cells
will "feel" lower concentrations.

The difference recorded for the effect of ethyleneglycol concentration on 0.1 mM ADD reduction by freely suspended and PAAH entrapped *Mycobacterium sp.* cells (Table 11) indicates that the protective effect recorded for yeast seems to be more general.

Table 11: Effect of EG on Δ^1-reduction of ADD (0.1 mM)[*] by freely suspended and PAAH entrapped *Mycobacterium sp.* cells.

EG content (%(v/v))	Relative activity (%)	
	free cells	PAAH entrapped cells
0	100	100
10	85	113
20	64	105
30	53	80
40	42	53

[*] ADD is water soluble at this concentration

In Conclusion

Criteria for cosolvent selection, based on solvent effects on enzyme stability and solvophobic interactions, was presented, discussed and tested on two cell models performing oxidation and reduction of steroids. Diols (e.g. ethyleneglycol) and DMSO in the concentration range 10-20% were found to be favourable solvents, while DMF, alcohols and others (e.g. acetone, dioxane) should be avoided. Appropriate cosolvent (e.g. ethylene glycol) when applied at optimal concentration allowed for increase in specific activity by the increased substrate solubility, although the cosolvent inhibited the enzymic activity. Protective effect exhibited by synthetic gel may allow for better use of cosolvents by reduction of the inhibitory effect.

164

References:

1. K. Kieslich. Steroid conversions, in Microbial Enzymes and Bioconversions. A.H. Rose (ed.) pp 370-465 (1980).

2. E. Antonioni, G. Carrea and P. Cremonesi. Enzyme catalyzed reactions in water-organic solvent two-phase systems. Enzyme Microb. Technol. 3, 291-296 (1981).

3. M.D. Lilly and J.M. Woodley. Biocatalytic reactions involving water-insoluble organic compounds, in Biocatalysts in Organic Synthesis. J. Tramper, H.C. van der Plas and P. Linko (eds.), Elsvier, pp 179-193 (1985).

4. L.E.S. Brink and J. Tramper. Optimization of organic solvent in multiphase biocatalysis. Biotech. Bioeng. 27, 1258-1269 (1985).

5. S. Fukui and A. Tanaka. Application of biocatalysts immobilized by prepolymer methods. Ad. Biochem. Eng. 29, 1-33 (1984).

6. K. Sonomoto, R. Matsuno, A. Tanaka and S. Fukui. Kinetic study on Δ^1 dehydrogenation of hydrocortisone by gel entrapped *Arthrobacter simplex* cells. J. Ferment. Technol. 62, 157-163 (1984).

7. K. Venkatsubramanian, A. Constantinides and W.R. Vieth. Synthesis of organic acids and modification of steroids by immobilized whole microbial cells. Enzyme Engineering 3, 29-41 (1978).

8. K. Sonomoto, A. Tanaka, T. Omata and S. Fukui. Application of photo-crosslinkable resin prepolymers to entrap microbial cells. Effects of increased cell-entrapping gel hydrophobicity on the hydrocortisone Δ^1-dehydrogenation. Eur. J. App. Microbiol. Biotechnol. 6, 325-334 (1979).

9. K. Sonomoto, I. Jin, A. Tanaka and S. Fukui. Application of urethane prepolymers to immobilization of biocatalysts: Δ^1 dehydrogenation of hydrocortisone by *Arthrobacter simplex* cells entrapped with urethane prepolymers. Agr. Biol. Chem. 44, 1119-1126 (1980).

10 K. Martinek and I.V. Berezin. General principles of enzyme stabilization. J. Solid-phase Biochem. 2, 343-385 (1978).

11.   J.F. Back, D. Oakenfull and M.B. Smith. Increa..ed thermal stability
      of proteins in the presence of sugars and polyols. Biochemistry 18,
      5191-5196 (1978).

12.   A. Ray. Solvophobic interactions and micelle formation in structure
      forming nonaqueous solvents. Nature 231, 313-315 (1971).

13.   T. Blank-Koblenc and A. Freeman. Effect of water miscible solvents
      on the kinetics and stability of glucose oxidase and pig-liver esterase.
      Manuscript in preparation.

14.   Q.H. Gibson, B.E.P. Swoboda and V. Massey. Kinetics and mechanism of
      action of glucose oxidase. J. Biol. Chem. 239, 3927-3934 (1964).

15.   A. Freeman, T. Blank and B. Haimovich. Gel entrapment of enzymes in
      crosslinked prepolymerized polyacrylamide-hydrazide. Annal. N.Y. Acad.
      Sci. 413, 557-559 (1983).

16.   C.O. Malikkides and R.H. Weiland. On the thermal denaturation of glucose
      oxidase. Biotech. Bioeng. 24, 1911-1914 (1982).

17.   T. Goren, M. Harnik, S. Rimon and Y. Aharonowitz. 1-Ene-steroid reductase
      of *Mycobacterium sp.* NRRL 3805. J. Steroid Biochem. 19, 1789-1797 (1983).

18.   I. Granot, Y. Aharonowitz and A. Freeman. Effect of water miscible solvents
      on Δ^1 reductase activity of free and PAAH entrapped *Mycobacterium sp.*
      Manuscript in preparation.

19.   A. Freeman and M.D. Lilly. Effect of water miscible solvents on Δ^1-dehydro-
      genase activity of free and PAAH entrapped *Arthrobacter simplex*. Submitted
      for publication.

20.   M. Popovich, H. Niebelschutz and M. Reuss. Oxygen solubility in fermentation
      fluids. Eur. J. App. Microbiol. Biotechnol. 8, 1-15 (1979).

21   A. Freeman. Gel entrapment of whole cells and enzymes in crosslinked,
      prepolymerized polyacrylamide-hydrazide. Annal. N.Y. Acad. Sci. 434,
      418-426 (1984).

22.   G. Pines and A. Freeman. Immobilization and characterization of *Saccharomyces
      cerevisiae* in crosslinked prepolymerized polyacrylamide-hydrazide. Eur. J.
      App. Microbiol. Biotechnol. 16, 75-80 (1982).

23.   Y. Dror and A. Freeman. Immobilization of *Saccharomyces cerevisiae* in
      polyacrylamide-hydrazide II: Effect of gel composition on tolerance
      to ethanol and other solvents. Manuscript in preparation.

IMMOBILIZED PLANT CELLS AS A SOURCE OF BIOCHEMICALS

Peter Brodelius

Institute of Biotechnology

Swiss Federal Institute of Technology

Hönggerberg, CH-8093 Zürich

Switzerland

Abstract. Plant cell culture methods are briefly reviewed. The biosynthetic capacity of cultivated plant cells and various approaches to increase productivity of such cultures are discussed. Techniques to immobilize large sensitive plant cells are described. The viability and bisynthetic capacity of immobilized plant cells are demonstrated with some model studies. Finally, the product release from immobilized plant cells is discussed.

INTRODUCTION

Pharmaceuticals, fragrances, flavors, and coloring principles from higher plants have been utilized by man for many years and these compounds continue to play an important role in modern technology. The majority of these substances are complex chemical structures and are classified as natural products. Over 80 % of the approximately 30,000 known natural compounds are of plant origin and, therefore, higher plants play an important role as a source of such substances. Most of these compounds cannot be produced by any other means (e.g. organic synthesis) than isolation from plant tissue. Since the supply of certain important raw plant materials is today or may in the near future be limited, it has become increasingly important to find and develop alternative means of production.

Twenty-five years ago it was suggested that plant cell cultures may be exploited as an alternative source of useful compounds [36]. It should in principle be possible to produce any compound found in the parent plant. Since then an increasing number of reports on the production of natural products by plant cell cultures have appeared [for review see e.g. 46,55].

167

PLANT CELL CULTURES

Various applications of plant tissue cultures are schematically illustrated in Figure 1. Plant cells in culture have been demonstrated to express "totipotency", which means that any living nucleated parendryma cell is capable of complete genetic expression, independent of its origin. Entire plants may therefore be regenerated from cultured cells. Morphogenesis, i.e. root and/or shoot formation, may be induced by altering the growth medium, in particular the hormone concentrations. Here the possible utilization of plant cell cultures for the production of biochemicals will be discussed. As indicated in Figure 1, the biosynthetic capacity of such cultures may be employed for bioconversions or for de novo synthesis from an inexpensive carbon source. A third approach is the feeding of precursors to the culture. This latter method requires knowledge about the biosynthetic pathway, as well as the availability of appropriate precursors.

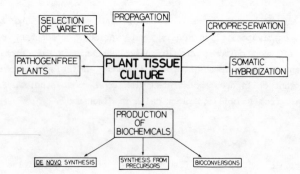

Figure 1. Schematic diagram of some selected applications of plant tissue cultures.

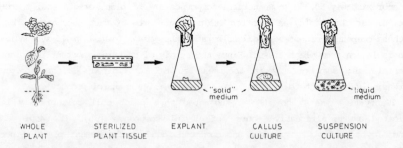

Figure 2. Schematic diagram of the principle procedure to establish a suspension culture from an intact plant.

From Plant To Suspension Culture

The standard procedure to establish a plant cell suspension culture is schematically shown in Figure 2. A piece of tissue (explant) is taken from the plant, and after surface sterilization it is placed on a medium solidified with agar. Various parts of the plant, such as root, stem or leave may serve as a source of the explant.

After the explant has been exposed to the medium for some time, a callus, which consists of meristematic cells, starts to form if the medium composition is appropriate. Newly formed callus tissue is transferred to fresh medium and this procedure is repeated until a more or less uniform tissue is obtained.

In the next step a friable callus tissue is transferred to a liquid medium and the suspension is incubated on a gyratory shaker. When required it is transferred to fresh medium to sustain growth.

Although whole plants have relatively simple requirements for growth, plant tissue cultures require a complex medium. The composition of the medium has to be empirically established for each plant species but in most cases a slight modification of a commonly used standard medium is sufficient for the successful establishment of a new culture.

The ingredients of a medium may be grouped into three classes, i.e. macro-, micro- and organic nutrients. The latter group includes vitamins and growth regulators (phytohormones) in addition to a carbon source (most commonly sucrose). Establishing and maintaining a culture is very much dependent on the type and concentration of the hormones used. The major classes of phytohormones are the auxins (e.g. indole-3-acetic acid (IAA), naphthalene acetic acid (NAA) and 2,4-dichlorophenoxy acetic acid (2,4-D)) and the cytokinins (e.g. kinetin and benzyladenine (BA)), which often are used in combination with one another.

PRODUCTION OF BIOCHEMICALS BY PLANT CELL CULTURES

The number of natural products isolated from plant cell cultures is large and this number is steadily increasing. From the concept of "totipotency" it should be possible to produce any compound present in the intact plant.

Selection Of High-Production Cell Lines

For the production of biochemicals using cultured plant cells a primary requirement is a good yield of the final product. The yield of a specific compound is generally low in culture. There are, however, methods available for establishing high-producing cell-lines. A number of examples have, indeed, been reported during recent years in which the yield on a dry weight basis is of the same order or even higher than that of the parent plant. Some of these examples are listed in Table 1.

The most widely used procedure to obtain high-producing plant cell cultures is based on the somaclonal variation within a cell culture. Single-cell clones are prepared and subsequently screened for productivity. Single cells are obtained from suspension cell aggregates by preparing protoplasts by enzymic digestion of the cell wall under hypertonic conditions. The protoplasts are plated at a low density and allowed to grow into callus cultures which subsequently are screened for productivity.

Another procedure involves selection of a callus culture with a comparably high content of the target substance [62]. This callus is divided into a number of pieces and subcultured. The selection and subculture procedures are repeated until a satisfactory production is reached.

An important requirement for an "industrial" plant cell culture is that it is stable in order to give a constant yield of product. However, the yield often declines with age of culture and reselection is frequently necessary.

Table 1. Some selected natural products formed in plant tissue cultures with a yield close to or higher than the parent plant [for references see 4].

Plant species	Compound	% of DW Culture	% of DW Plant	type of culture*	g/l culture
Panax ginseng	ginsengoside	27	4.1	C	-
Morinda citrifolia	anthraquinone	18	2.2	S	2.5
Vitis sp.	anthocyanins	16	n.s	S	0.83
Coleus blumei	rosmarinic acid	15	3.0	S	3.6
Lithospermum erythrorhizon	shikonin	14	1.5	S	1.5
Coptis japonica	benzylisoquino- line alkaloids	11	5-10	S	1.7
Coptis japonica	berberine	10	2-4	S	1.2
Galium mollugo	shikimic acid	10	n.s	S	1.2
Nicotiana tabacum	cinnamoyl putre- scine	10	n.s	S	1.0
Berberis stolonifera	jatrorrhizine	7	n.s	S	2.7
Glycyrrhiza urarensis	glycyrrhizin	7	n.s	C	-
Cassia tora	anthraquinones	6	0.6	C	-
Nicotiana tabacum	nicotine	3.4	2.0	C	-
Stephania cepharantha	biscoclaurine	2.3	0.8	C	-
Panax ginseng	saponins	2.2	0.5	S	n.s
Dioscorea deltoides	diosgenin	2	2.0	S	n.s
Peganum harmala	harmane	2	n.s	S	0.12
Coffea arabica	caffeine	1.6	1.6	C	-
Fedia cornucopiae	valtrate	1.4	0.6	S	n.s
Catharanthus roseus	ajmalicine	1.0	0.3	S	0.26
Catharanthus roseus	serpentine	0.8	0.5	S	0.16
Nicotiana tabacum	ubiquinone-10	0.5	0.003	S	n.s
Tripterygium wilfordii	tripdiolide	0.05	0.001	S	0.004

(*) C = callus culture; S = suspension culture; n.s = not stated

Table 2. A comparison of some characteristics of plant and microbial cells.

	Microbial cell	Plant cell
Size	~2 u^3	>10^5 u^3
Shear	intensitive	sensitive
Water content	~75 %	~90 %
Doubling time	1 H	days
Oxygen consumption	1-2 vvm	0.3 vvm
Fermentation time	days	weeks (2-3)
Product	extracellular	intracellular
Media cost	~6 $/m3	~50 $/m3
Mutation	possible	require haploid cells

Large Scale Cultivation Of Plant Cells

The technology to cultivate microorganisms in large fermenters has during recent years been extensively developed. A number of sophisticated processes are in operation. This know-how cannot, however, be directly transferred to the cultivation of plant cells on a large scale since cultivated plant cells and microorganisms differ in many respects as outlined in Table 2. Fermenters developed for the cultivation of microorganisms cannot in most cases be directly employed for growth of plant cells.

An industrial fermentation of plant cells would result in certain advantages over field cultivation or collection of plant material for isolation of high value chemicals as summarized in Table 3.

Production under controlled conditions. Plant cells in culture may be grown under controlled conditions in well-defined media in contrast to field cultivation of plants which often are exposed to non-controllable factors such as climate and parasites. The medium used for the cultivation can be optimized for either growth or product formation.

Increased Productivity. Cultured plant cells may be manipulated to produce increased amounts of the target substance by optimization of the medium or by addition of inducers to the medium. In this respect microbial elicitors, which induce secondary metabolism in whole plants, have received increasing attention.

Table 3. Some advantages of plant cell cultures over field cultivation or collection of plant material.

- Production under controlled conditions
- Manipulation of the cells for increased productivity
- Constant supply of raw materials
- New compounds

172

Table 4. Some examples of secondary products that have been induced in plant cell cultures by treatment with microbial elicitors.

Plant species	Microbial species*	Product	Concentration (mg/g dry weight)		Ref.
			before	after	
Canavalia ensiformia	Pithomyces chartarm	medicarpin	0	0.42	[28]
Cephalotaxus harringtonia	Vericillium dahliae	harringtonine alkaloids	0.01	0.51	[30]
Cinchona ledgriana	Aspergillus niger	anthraquinones	3	15	[59]
Dioscorea deltoides	Rhizopus arrhizus	diosgenin	25	72	[50]
Glycine max	Phytophthora megasperma	glyceollim	0	0.05	[17]
Glycine max	Saccharomyces cerevisiae	glyceollin	0	0.20	[24]
Papaver somniferum	Verticillium dahliae	morphine	0.07	1.25	[30]
		codeine	0.08	1.11	
Ruta graveolens	Rhodotorula rubra	rutacridon-epoxides	0	0.23	[18]
Thalicrum rugosum	Saccharomyces cerevisiae	berberine	20	50	[24]

*Source of elicitor

The effects of various elicitors on secondary metabolism in cultivated plant cells are exemplified in Table 4.

To demonstrate this approach to increase the productivity of a plant cell suspension culture, one example will be given. A carbohydrate fraction isolated from yeast extract can be used as an elicitor to induce an increased synthesis of the benzylisoquinoline-alkaloid berberine in suspension cultures of Thalictrum rugosum [24]. A typical growth curve of the Thalictrum suspension culture is shown in Figure 3. The cells produce berberine during growth but when the carbon

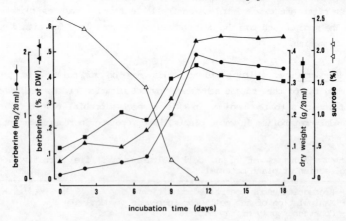

Figure 3. Growth curve and berberine synthesis of a suspension culture of Thalictrum rugosum [from 24].

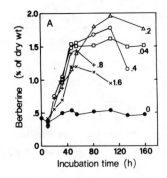

Figure 4.
Berberine synthesis in suspension cultures of Thalictrum rugosum as function of incubation time after treatment with various concentrations of yeast elicitor [from 24].

source (sucrose) is consumed both growth and alkaloid synthesis cease. However, if the suspension culture is treated with elicitor at this stage more berberine is synthesized. The response of the culture is dependent on elicitor concentration during treatment as illustrated in Figure 4. At relatively high elicitor concentrations, a reduced berberine synthesis is observed, most probably due to cell death. Consequently, an optimal elicitor concentration may be determined.

The effect of the elicitor on berberine synthesis is also influenced by the physiological status of the cells. When the elicitor is added to a rapidly growing suspension culture (early exponential phase) the observed effect on berberine synthesis is relatively limited. However, if the elicitor is added in late exponential or in stationary phase, a pronounced increase in berberine synthesis is obtained.

Other stress inducers (e.g. salt and light) may be employed for a similar induced/increased synthesis of the target substance. While manipulations of this kind are difficult to carry out on intact plants, this approach to increase productivity is limited to cell cultures.

Constant supply of raw material. In vitro cultivation of plant cells would guarantee a constant supply of raw material under batch or continuous culture conditions. No seasonal variations would occur and no risk for loss of a crop due to environmental reasons (e.g. parasites or draught) would exist. Furthermore, production of phytochemicals in a fermentor would eliminate any possible shortage or sudden price increases due to political instabilities.

New compounds. Completely new compounds not present in the plant have been found

Table 5. Some factors limiting the industrial utilization of plant cell cultures
----
                    - Low yield of product
                    - Slow growth of cells
                    - Genetic instability of isolated cell lines
                    - Cell organisation and differentiation
                    - Cell aggregation
                    - Product release
----

in cell cultures. It should also be possible to produce new compounds by feeding analogues of natural intermediates of a biosynthetic pathway. These analogues may be enzymatically incorporated or converted into new compounds.

Limitations. There are numerous reasons why progress has been slow in the large scale industrial application of plant cell cultures for the production of biochemicals. Table 5 lists some of the limiting problems.

Growth characteristics of cells. Plant cells grow slowly in culture with growth rates corresponding to a doubling time of 20-60 hours. This slow growth rate makes the generation of biomass for isolation of biochemicals prohibitively expensive unless the culture produces large amounts of a high value substance. Furthermore, the slow growth rate complicates scale-up since a relatively large reactor volume is required and extreme precautions must be taken against external sources of contaminations.

Genetic instability of cells. There are a number of methods available for the selection of high-producing cell lines as described above. The capacity of these selected cell lines to produce the target substance often declines upon serial subcultivations. For an industrial application of plant cell cultures a very important requirement is a stable and constant product yield.

Cell aggregation and mass transfer effects. Plant cells in suspenion culture most often grow in aggregates of various sizes. Cells in the center of such a macroscopic aggregate are exposed to a different environment than are cells on the surface because of mass transfer limitations. It can therefore be expected that cells in different parts of the same aggregates behave differently leading to complications in optimizing product formation.
    The wide size distribution of aggregates in a culture may impose problems in the scale-up process since the physical-mechanical conditions in a large fermenter differs from those in a small vessel. It is possible that the different settling rates of aggregates coupled with non-uniformities in fluid mixing in a large fermenter will further increase the range of aggregate sizes and, thereby, increase the heterogenity of the culture.

Cell organization and differentiation. It has in a number of cases been suggested that organization and/or differentiation of the cultivated cells is required for product formation. The induction of differentiation and cultivation of differentiated cells may be too complicated to carry out on an industrial scale.

Product release, recovery and purification. The recovery of the target substance may be difficult because yields obtained from plant cell cultures are often low. This can be due either to a low productivity or the extraction method used.

Low shear resistance. Cultivated plant cells are relatively large and therefore often sensitive to shear forces. Cultivation of these sensitive cells on a large scale at high cell densities impose considerable problems in nutrient and oxygen supply. In most cases it can be expected that specially designed fermenters have to be used. The air-lift fermenter has been shown to be appropriate for the cultivation of sensitive plant cells.

THE FIRST INDUSTRIAL PROCESS

Despite the problems encountered in the scale-up process, the first chemical compound from a plant cell culture was introduced on the market recently. The compound, shikonin, is a red dye and also a pharmaceutical used in Japan traditionally for its antibacterial and anti-inflammatory effects. It is produced from suspension culture of Lithospermum erythrorhizon by Mitsui Petrochemical Industries Ltd. in a two-stage batch process according to Figure 5. When this

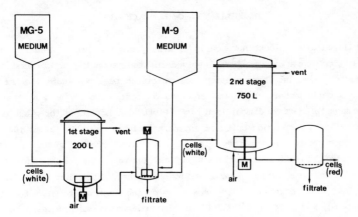

Figure 5. Schematic flow diagram of the two-stage process used for production of shikonin by cultivation of Lithospermum erythrorhizon. Note: the inoculated cells and the cells obtained after the first stage are white while the cells obtained after the second stage are red from the shikonin produced.

Table 6. A comparison of shikonin production from intact plants and cultivated plant cells [from 5].

	Time before harvest	Shikonin (% of dry weight)
Intact plant	2-3 years	1-2
Cultivated cells	3 weeks	14

process was established, the product was sold for around $4,000 per kilogram, which well exceeds the $1,000 per kilogram that has been estimated as the lowest price of a compound worth producing by plant cell culture at the current levels of technology. The cells are cultured for three weeks and from published values a yield of about 1.2 kg per batch can be estimated for the pilot plant process outlined in Figure 5 [21,22]. This new process has opened new areas for utilization of shikonin. In Japan "biocosmetics" have become popular during recent years and shikonin is used as a dye in such preparations (e.g. biolipstick).

This process may be considered as a major breakthrough in plant cell culture technology. It has been shown that a natural product may be produced by plant cell culture techniques on a commercial basis. A comparison between culture and plant (root) as summarized in Table 6 shows some distinct advantages of the new process. The high productivity in culture (1.5 g/l) has been achieved after extensive optimization work. The overall productivity in culture has been improved 13-fold by the two-stage procedure. This clearly demonstrates the power of decoupling production from growth.

IMMOBILIZATION OF PLANT CELLS

Immobilized biocatalysts have during recent years received increasing attention. For some years we have been involved in studies on the immobilization of cultivated plant cells. These studies were initiated to investigate whether such immobilized cells retain viability and biosynthetic capacity and whether the inherent advantages of an immobilized biocatalyst also hold true for plant cells.

Plant cells in suspension cultures are relatively large cells (up to 100 um in diameter; see Table 2) and they grow, as mentioned above, in aggregates of various sizes. These characteristics of cultivated plant cells have to be considered when the method of immobilization is chosen. Entrapment within various gels or membrane reactors is the most appropriate method to use for these heterogeneous cell preparations. Other immobilization techniques, such as covalent attachment and adsorption, have also been employed.

Entrapment.

Alginate. In the very first investigations of immobilized plant cells, entrapment in calcium alginate was used [13]. Alginate was selected as being a very mild and simple method of immobilization which can be carried out under sterile conditions.

Agar and agarose. Agar and agarose form gels upon cooling a heated aqueous solution. The gelling temperature can be altered by chemical modification of the structure (introduction of hydroxyethyl groups).

For the immobilization of plant cells a gelling temperature around 30 °C is appropriate. A convenient method to make spherical beads of these polymers has recently been developed [47]. This general procedure for the immobilization of cells with preserved viability is based on dispersion of the cell/polymer suspension in a non-toxic hydrophobic phase (e.g. vegatable oil) under stirring. When droplets of appropriate size have been obtained the whole mixture is cooled to allow gel formation. The size of the beads can be controlled by the stirring speed.

Kappa-carrageenan. Kappa-carrageenan is a strong anionic hydrocolloid which may be used for the entrapment of cells under relatively mild conditions in a manner similar to that for alginate. The polymer solution must, however, be heated in order to be maintained in a liquid state. Beads are formed by dripping the cell/carrageenan suspension into a medium containing potassium ions (0.3 M). Gel particles (beads) may also be prepared in analogy with agarose beads in a two-phase system. In this case the gel particles are treated with a potassium solution to improve the mechanical stability. Most plant cell media contain sufficient amounts of potassium and, therefore, no extra additions of potassium are required.

Polyacrylamide. Polyacrylamide has been extensively used for the immobilization of microbial cells [9]. Attempts to entrap plant cells in this polymer resulted in non-viable cell preparations [7]. The reason for this was the toxicity of the chemicals used to prepare the polyacrylamide gel.

Plant cells have, however, been immobilized in polyacrylamide with preserved viability according to a newly developed and elegant technique [20]. In this method the plant cells are suspended in an aqueous solution of prepolymerized linear polyacrylamide, partially substituted with acylhydrazide functional groups. Subsequently, the suspension is cross-linked with a controlled amount of dialdehyde (e.g. glyoxal) under physiological conditions. The gel obtained is mechanically disintegrated into smaller particles.

Polyurethane. Polyurethane foam particles were used to immobilize plant cells [38]. Freely suspended cells were found to invade such porous particles where they were strongly retained. The mechanism responsible for the immobilization is not quite clear but it is suggested that the cells are entrapped within the foam particles.

Membrane reactors. Various membrane configurations may also be used for the entrapment of plant cells. Hollow-fiber reactors were the first examples of this type of immobilization [35,53]. The plant cells are introduced on the shell side of the reactor and oxygenated medium is supplied at a relatively high flow rate through the fiber lumen. It has been shown in other types of reactors that the pore size of the membrane is of great importance for cell behavior [54].

Adsorption.

Adsorption to a solid support offers a simple and mild method to immobilize cells [9]. The relatively weak binding forces involved will, however, most likely lead to release of cells from the support, especially if the cells are allowed to grow and divide in situ. One example of adsorbed plant cells in the form of protoplasts has been reported [3]. The protoplasts were adsorbed on microcarriers (Cytodex), widely used for growth of anchorage-dependent animal cells.

Covalent linkage.

Covalent linkage involves, as the name implies, covalent bonds between the polymer and the biocatalyst. The introduction of such bonds require reactive chemicals and may result in loss of cell viability. Therefore, this method is less attractive for the immobilization of sensitive plant cells.

One example can be found in the literature. Cells of Solanum aviculare were covalently linked to glutaraldehyde activated polyphenylene oxide beads [32]. The immobilzed cells were biosynthetically active but no information on cell viability was given.

VIABILITY OF IMMOBILIZED PLANT CELLS

The entrapment methods described above have been used for immobilization of various plant cells as summarized in Table 7.

It is of great importance that the plant cells remain viable after immobilization, since major parts of the cell metabolism are required for the conversion of an inexpensive carbon source to complex molecules. As discussed above the slow growth of cells is a major limitation in suspension cultures (doubling time is

179

Table 7. Viable preparations of plant cells entrapped in various matrices.

Matrix	Plant species
Alginate	Cannabis sativa [34]; Catharanthus roseus [7]; Coffea arabica [29]; Daucus carota [33]; Digitalis lanata [1,13]; Glycine max [47]; Ipomoea sp. [34]; Lavendula vera [45]; Morinda citrifolia [13]; Mucuna pruriens [58]; Nicotiana tabacum [43]; Papaver somniferum [25]; Silybum marianum [16].
Agarose	Catharanthus roseus [7]; Daucus carota [8]; Datura innoxia [8]; Glycine max [47].
Agar	Catharanthus roseus [7]; Daucus carota [39]; Glycine max [47].
Carrageenan	Catharanthus roseus [7]; Daucus carota [39].
Polyacrylamide	Catharanthus roseus [51]; Nicotiana tabacum [51]; Mentha-species [27].
Polyurethane	Capsicum frutescens [38].
Membrane	Glycine max [53]; Daucus carota [35].

normally 20-60 hours). On the contrary with immobilized cell systems, the growth of cells is actually a problem. Extensive growth will in this case result in cell release and even destruction of beads. Therefore, at present, attempts are being made to formulate media which keep the cells in a viable state over an extended period of time and at the same time do not allow multiplication of the cells. However, resting viable cells may be induced to grow in situ by administration of a growth medium when the biosynthetic capacity of the immobilized preparation is detoriating, leading to an increased operational life of the biocatalyst. Various techniques, both direct and indirect, may be employed to investigate the viability of immobilized plant cells.

Respiration.

Respiration of cells can be used as an indication of cell viability. This is particularly true when the respiration is measured over an extended period of time. However, as the cell density increases within the beads due to cell growth the measured relative respiration decreases as illustrated in Figure 6. For

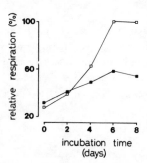

Figure 6.
Relative respiration of Catharanthus roseus cells. (-■-) Cells immobilized in alginate; (-□-) cells after dissolving the polymer.

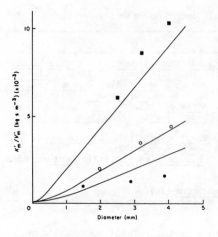

Figure 7.
Slopes of Lineweaver-Burke plots for oxygen consumption by alginate-entrapped cells of Daucus carota as function of bead diameter. The symbols represent experimental values and the solid lines theoretical values. Cell loadings were 50.6 (—■—), 14.1 (—O—) and 5.6 (—●—) % (w/w) [from 31].

quantitative determination of cell respiration the cells should be set free.

The influence of support material, cell loading and bead diameter on the rate of repiration of immobilized Daucus carota cells was investigated [31]. The respiration kinetics of alginate-entrapped cells in the form of Lineweaver-Burke plots were determined for various cell loadings and bead diameters. The slope of the plots become steeper with increasing cell density or bead size indicating decreasing respiration rate as a result of oxygen diffusion limitations. In oFigure 7 the slopes of the plots are plotted as a function of bead size for three different cell loadings. The theoretical curves were obtained by a computer simulation. The deviation of these curves at high and low cell densities may be explained by the fact that the effective diffusion coefficient is affected by the cell loading.

The maximum bead diameter for unlimited respiration (ϕ_{crit}) was derived from the generalized Thiele-modulus [31]:

$$(\phi_{crit})^2 = 24\ D_e\ \frac{(C_s - C_{eq})}{r_s} \qquad (1)$$

where r_s=oxygen consumption rate; C_s=oxygen concentration at bead surface; C_{eq}=equilibrium concentration of the oxygen consuming reaction; D_e=effective diffusion coefficient of oxygen within the gel matrix. Substituting the experimental data in this equation yields the results summarized in Table 8.

Assuming that the oxygen concentration at the bead surface will be about 95 %

Table 8. Critical bead diameter (in mm) for alginate-entrapped cells of <u>Daucus</u>
<u>carota</u> calculated with equation (1) [from 31].

Cell loading	Surface concentration of oxygen (% saturation)				
(% w/w)	2	20	50	75	95
5.6	0.55	1.74	2.76	3.38	3.80
14.1	0.39	1.23	1.96	2.39	2.69
50.6	0.20	0.64	1.02	1.25	1.40

of saturation, one can see from Table 8 that for good diffusion characteristics
the bead diameter may be as high as 1.4 and 3.8 mm for cell densities of 50 and 5
% (w/w), respectively.

Oxygen concentration profiles within a bead may be calculated by means of a
numerical solution of a differential mass balance over the bead with internal
oxygen diffusion and consumption. Figure 8 shows such calculated oxygen concentration profiles.

The influence of bead diameter on the extent of respiration by D. carota cells
entrapped in various polymers was also investigated [31]. Evaluation of the data
led to the conclusion that only minor differences exist between the various gel
materials. Consequently, diffusion limitations of oxygen in alginate is not
likely to be the reason for the stimulated secondary product formation observed
for cells entrapped in this polymer (see below).

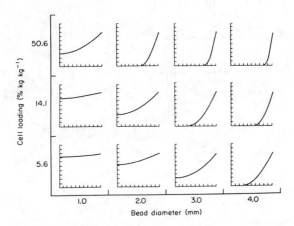

Figure 8. Calculated oxygen concentration profiles in alginate beads as a
function of bead diameter and cell loadings. Range x-axis: from centre
(left) to surface (right) of the bead. Range y-axis: from zero (bottom)
to 0.3 mM (top) [from 31].

Cell Growth And Division.

Cell growth [12,47] and cell division [12] are proofs of cell viability. These events may be studied by following the increase of cell number (cell density) or dry weight within the beads or by studying the mitotic index which reflects the number of cells in mitosis.

In Vivo NMR.

Phosphorus-NMR has emerged in the last decade as a powerful technique for the study of intracellular metabolism [26]. Although relatively few studies dealing with cultivated plant cells have been reported, it has become clear that the levels of the major phosphorylated metabolites, such as ATP, can be monitored with this non-invasive method [10,11,61]. Moreover, since the chemical shift of various resonances are a function of pH, the cytoplasmic and vacuolar pH-values are directly obtained from the spectra [41,49]. Furthermore, uptake and storage of inorganic phosphate (P_i) in different intracellular compartments can be probed [11].

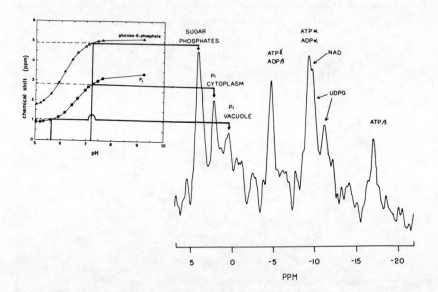

Figure 9. A typical ^{31}P NMR spectrum (145.7 MHz; 40000 scans) of C. roseus cells. The inserted figure shows the chemical shift of inorganic phosphate and glucose-6-phosphate as function of pH [from 11,57].

Figure 9 depicts a typical ^{31}P NMR spectrum obtained for an oxygenated sample of freely suspended C. roseus cells. Two different P_i resonances are observed at chemical shifts corresponding to a pH of 7.3 (cytoplasma) and 5.7 or below (vacuole), respectively (as indicated in the inserted figure). As discussed above entrapment of the cells in a polymer may prevent the proper oxygenation of the cells. The effects of anaerobic conditions on the NMR measurable parameters of freely suspended cells are shown in Figure 10. No changes are observed in the vacuolar pH or P_i-level. However, drastic changes are seen in the ratio of the sugar phosphate and the cytoplasmic P_i resonances. Moreover, the upfield shift of both these resonances indicates that the cytoplasmic compartment has been acidified to about pH 6.8 during anaerobic conditions. Upon reperfusion with oxygenated buffer the intracellular pH goes back to pH 7.3. The acidification during hypoxia arises from an increase in the lactate content and is possibly also due to an increase of carbondioxide within the cells.

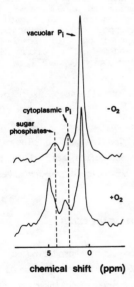

vacuolar P_i

cytoplasmic P_i

sugar phosphates

-O_2

+O_2

5   0

chemical shift (ppm)

Figure 10.
A comparison of the downfield region of the ^{31}P NMR spectrum of C. roseus cells under aerobic and anaerobic conditions [from 57].

Figure 11 compares the ^{31}P NMR spectra of freely suspended and agarose- and alginate-entrapped cells under oxygenated conditions. It is immediately apparent that the same resonances are observed in all three spectra and the chemical shifts of the cytoplasmic sugar phosphate and P_i resonances are almost identical (4.9 and 3.0 ppm, respectively). Thus, in most cells the intracellular pH of the cytoplasmic compartment is maintained at pH 7.3. Also the ADP/ATP ratio calculated for the three spectra shown in Figure 11 is the same within experimental error indicating a similar energy status for the three cell preparations.

184

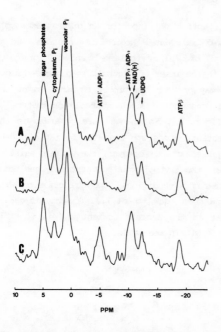

Figure 11.

^{31}P NMR spectra obtained for various preparations of C. roseus cells. (A) alginate-entrapped cells after 72 h incubation; (B) agarose-entrapped cells after 72 h incubation; (C) freely suspended cells after 48 h incubation [from 57].

Although it is well known that changes in the intracellular pH exert regulatory effects on the metabolism of plant cells it is clear from these NMR studies that the cytoplasmic pH, as well as the vacuolar pH, are not altered by the entrapment of the cells in agarose or alginate. The changes in biosynthetic behavior that are sometimes observed for immobilized cells are likely to be caused by more subtle changes than those that can be followed non-invasively by ^{31}P NMR.

BIOSYNTHETIC CAPACITY OF IMMOBILIZED PLANT CELLS

Figure 12 demonstrates various alternative routes from a high-producing cell line to product. Biomass is generated by cultivating the cells on a large scale in a medium optimized for growth. When the product formation is linked to growth (growth associated products) the biomass is extracted and the product isolated. For non-growth associated products a different approach should be employed. The biomass is transferred to a second cultivation stage where the conditions are optimized for product formation. This step may be carried out in free suspension or in the immobilized state. Two different process designs are possible for immobilized cells depending on whether the product is extra- or intracellular.

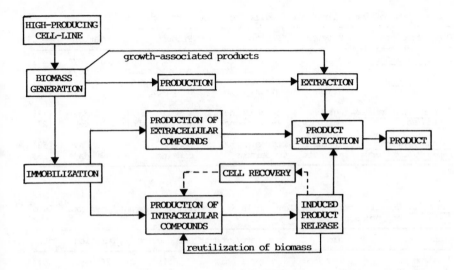

Figure 12. Flow sheet of various alternative routes to produce biochemicals with a high-producing plant cell culture system.

In the former case a continuous process may be used and in the latter a semi-continuous process with intermittant release of product may be employed.

Immobilized plant cells have been utilized for a wide range of biosynthetic studies including bioconversions, synthesis from added precursors and de novo synthesis from an inexpensive carbon source. In the following these investigations will be briefly reviewed.

Bioconversions.

Plant cells in culture have been utilized for a wide range of bioconversions [48]. These include hydroxylations, methylations, acetylations and glycosylations. Immobilized plant cells have been employed for bioconversions as outlined in Table 9.

Synthesis From Added Precursors.

The yield of a particular compound in plant cell cultures may be increased considerably by feeding appropriate precursors. Although this approach to synthesize complex biochemicals appear to be attractive relatively few examples can be found in the literature. Table 10 lists precursor feedings to immobilized plant cells.

186

Table 9. Use of immobilized plant cells for bioconversions.

Plant species	Polymer	Substrate	Product	Duration	Ref.
Digitalis lanata	alginate	digitoxin	digoxin	70 days	[15]
	alginate	methyldigitoxin	methyldigoxin	180 days	[2]
Daucus carota	alginate	digitoxigenin	periplogenin	12 days	[33]
	alginate	gitoxigenin	5-hydroxygi-toxigenin	21 days	[56]
Catharanthus roseus	agarose	cathenamine	ajmalicine	1 day	[19]
Mentha sp.	polyacryl-amide	(−)menthone	(+)neomenthol	1 day	[27]
Mucuna pruriens	alginate	tyrosin	DOPA	1 day	[58]
Papaver somniferum	alginate	codeinone	codeine	30 days	[25]

Table 10. Use of immobilized plant cells for converting distant precursors into end products.

Plant species	Polymer	Substrate/Product	Operation	Duration	Ref.
Capsicum frutescens	polyuretane foam	isocapric acid capsaicin	recirculated column	10 days	[38]
Catharanthus roseus	alginate, agarose, agar, carrageenan	tryptamine + seco-loganin ajmalicine isomers	batch	5 days	[7]
Catharanthus roseus	alginate, agarose	tryptamine + seco-loganin ajmalicine isomers	batch change	14 days	[8]

De Novo Synthesis

A wide varity of compounds have been isolated from plant tissue cultures [46,55]. A considerable interest has recently been shown for the production of

Table 11. Use of immobilized plant cells for de novo synthesis.

Plant species	Polymer	Product	Operation	Duration	Ref.
Capsicum frutescens	polyurethane	capsaicin	batch	10 days	[38]
Catharanthus roseus	alginate, agarose, agar, carrageenan	ajmalicine	batch	14 days	[8]
Catharanthus roseus	alginate	serpentine	batch	40 days	[15]
Catharanthus roseus	alginate + polyacrylamide	serpentine	batch change	220 days	[37]
Coffea arabica	alginate	methylxanthines	batch	32 days	[29]
Glycine max	hollow fibers	phenolics	continuous	30 days	[53]
Lavandula vera	alginate	blue pigment	batch change	200 days	[45]
Morinda citrifolia	alginate	anthraquinones	batch	23 days	[14]
Solanum aviculare	polyphenylene-oxide	steroid glyco-sides	continuous	11 days	[32]

Table 12. Examples of alginate-entrapped plant cells producing higher amounts of secondary products than freely suspended cells.

Plant species	Product	Incubation time	Ratio*	Ref.
Catharanthus roseus	ajmalicine	12 days	1.4	[7]
Catharanthus roseus	serpentine	40 days	4	[15]
Coffea arabica	methylxanthines	32 days	4.5	[29]
Morinda citrifolia	anthraquinones	23 days	10	[14]

*immobilized/free

such compounds by immobilized plant cells. A number of model studies have been carried out as summarized in Table 11 in order to investigate whether immobilized cells behave different from freely suspended cells in this context. The carbon source was sucrose in the listed examples.

Alginate-entrapped cells have in serveral instances shown an enhanced product formation as compared to freely suspended cells under the same conditions as summarized in Table 12. Here the synthesis of methylxanthine alkaloids by immobilized cells of Coffea arabica will be used to illustrate this phenomenon [29].

The productivity of methylxanthine alkaloids (hypoxanthine and coffeine) by suspension cultured and alginate-entrapped cells of C. arabica is depicted in Figure 13. Significant synthesis of alkaloids did not start until the sucrose in the medium had been consumed and growth had ceased.

Further experiments were carried out to investigate the effects of alginate-entrapment on alkaloid biosynthesis in C. arabica. Free and immobilized cells

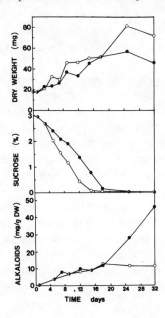

Figure 13.
Dry weight increase, sucrose consumption and methylxanthine alkaloid production of C. arabica cells as function of incubation time. (—O—) Freely suspended cell; (—●—) immobilized cells [from 29].

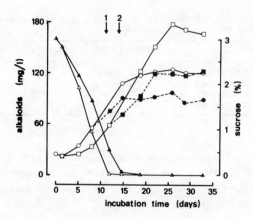

Figure 14. Sucrose consumption and alkaloid synthesis in cells of C. arabica as funcion of incubation time. (-O-) suspended cells; (-●-) suspended cells immobilized at day 11 (arrow no. 1); (-□-) immobilized cells; (-■-) immobilized cells set free on day 14 (arrow no. 2); (-△-) and (-▲-) sucrose concentration for free and immobilized cells, respectively [from 29].

were incubated in multiple samples under identical conditions. When sucrose was depleted (day 11) the cells from half of the suspension cultures were immobilized in alginate and subsequently returned to the depleted media. Likewise when no sugar remained in the immobilized preparations (day 14) half of the preparations were treated with citrate buffer to release the cells. The released cells were washed and transferred back to the original media. The four different set of preparations so obtained were cultivated further and the results are depicted in Figure 14.

The cells immobilized throughout the experiment produced more alkaloids than the other three preparations. Entrapment of cells after depletion of sugar had a negative effect on alkaloid synthesis. Apparently, energy is required for the cells to adapt to the immobilized state. On the other hand when the matrix was dissolved and the cells set free, the alkaloid synthesis continued at a constant rate. However, the synthesis stopped when the concentration of alkaloids had reached the same level as that of cells grown in suspension throughout the experiment.

From these experiments it may be concluded that the immobilization does not change the cell metabolism irreversibly. The increased alkaloid synthesis by immobilized cells may be explained by a reversible interaction between cells and alginate and/or gradients of nutrients/products (diffusion barriers) within the preparation. Various studies have, however, shown that diffusion limitations are not so pronounced in slow-growing immobilized plant cell preparations. Indications of this are the observations that immobilized cells are appropriately oxygenated [31,57] and that the phosphate metabolism is essentially unaffected by immobilization [57]. On the other hand reversible interaction between alginate and pectic acid (a component of the cell wall) could be established by calcium ions. In this way alginate could function as a glue between cells and thereby mediate a cell to cell interaction mimicing that of a differentiated plant tissue. Of the various polymers used for the entrapment of plant cells only alginate can function in this manner and it is only in alginate-entrapped cells that an increased secondary product formation has been observed [4].

An alternative or additional explanation to the increased production may be found in the suggestion that alginate can function as an elicitor (see above) in plant cell systems [60]. In fact, we have observed that alginate can induce phenylalanine ammonia lyase in suspension cultures of soybean (<u>Glycine</u> <u>max</u>) in a similar manner as the glucan elicitors [23]. However, in this case of alginate no phytoalexin (glyceollin isomers) was synthesized indicating that alginate elicits another biosynthetic pathway in soybean cells than the glucan elicitor.

PRODUCT RELEASE FROM IMMOBILIZED PLANT CELLS.

A major advantage of immobilized biocatalysts is the possible continuous operation of a bioreactor. This distinct advantage can, however, not be realized with immobilized plant cells since secondary products most often are stored within the vacuoles of such cells. Attempts are therefore being made to induce product release from immobilized cells.

Spontaneous release from immobilized plant of products, which are normally stored within the cells in suspension, has been reported [15,52]. However, in most cases product release has to be induced by external manipulation of the cells. This induced product release should preferable not affect the viability of the cells. In order to release products from vacuoles of immobilized cells two membrane barriers have to be penetrated (i.e. the plasma membrane and the tonoplast surrounding the vacuoles). The permeability of these membranes may be monitored in various ways as described below. Different chemical substances, as exemplified in Table 13, may be employed to make membranes within plant cells permeable to various compounds.

<u>Plasma Membrane.</u> Various enzyme activities within the cells can be employed for monitoring the permeability of the plasma membrane. Enzymes requiring nucleotide coenzymes such as NADP(H), ATP or CoA are particularly convenient to use. These coenzymes cannot penetrate an intact plasma membrane and, therefore, no enzyme activity is expressed unless the plasma membrane has been made permeable [19].

Table 13. Some examples of chemical substances that may be employed for the permeabilization of plant cells.

----
dimethylsulfoxide	chloroform	toluene
phenethyl alcohol	protamine sulfate	poly-L-lysine
nystatin	cytochrome c	lysolecithin
triton X-100	hexadecyltrimethylammonium bromide (HDTMAB)	

The expressed activity of an intracellular enzyme (isocitrate dehydrogenase) is dependent on concentration of permeabilizing agent as summarized in Table 14 for three different plant cell cultures.

Table 14. Concentration of various permeabilizing agents required for expression of 50 and 90 % of maximum isocitrate dehydrogenase activity in cultivated plant cells [from 6].

Permeabilizing agent	Catharanthus roseus		Chenopodium rubrum		Thalictrum rugosum	
	50%	90%	50%	90%	50%	90%
DMSO (% v/v)	2	5	5	30	7	15
PEA (% v/v)	0.82	0.96	0.66	0.86	0.60	0.72
Chloroform (% sat.)	58	68	42	53	55	70
Triton X-100 (ppm)	140	190	120	145	115	200
HDTMAB (ppm)	12	28	12	24	13	28

Tonoplast. To monitor the permeability of the tonoplast (membrane surrounding the vacuole) various compounds stored in the vacuoles may be used. ^{31}P NMR spectroscopy has been used to study the permeability of tonoplasts in Catharanthus roseus cells [40]. These cells accumulate inorganic phosphate in the vacuoles and after treatment with various concentrations of DMSO the size of the vacuolar phosphate peak is reduced as illustrated in Figure 15.

In other instances the direct measurement of the release of secondary products from the vacuoles may be used. The release of betanin and berberine from cells of Chenopodium rubrum and Thalicrum rugosum, respectively, after treatment with various permeabilizing agents, is summarized in Table 15.

Table 15. Concentration of various permeabilizing agents required for release of 50 and 90 % of intracellularly stored products from cultivated plant cells [from 6].

Permeabilizing agent	Catharanthus roseus		Chenopodium rubrum		Thalictrum rugosum	
	50%	90%	50%	90%	50%	90%
DMSO (% v/v)	3	7	10	35	13	30
PEA (% v/v)	1.04	1.16	0.86	0.98	0.60	0.80
Chloroform (% sat.)	53	66	54	64	50	67
Triton X-100 (ppm)	140	185	185	230	140	210
HDTMAB (ppm)	44	72	22	84	24	60

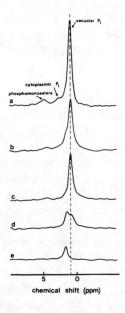

chemical shift (ppm)

Figure 15.
Downfield region of the ^{31}P NMR spectra of Catharanthus roseus cells after treatment with various concentrations of DMSO. (A) Untreated cells; (B) 1 %; (C) 2 %; (D) 5 %; (E) 10 % DMSO (v/v). [from 40].

<u>Viability of permeaaabilized cells.</u> Cell viability after treatment with the permeabilizing agent is a requirement for the application of intermittant release of intracellularly stored products in a process. Studies have, however, shown that cells are adversely affected by the permeabilization with various chemicals. Cells of Chenoposium rubrum and Thalicrum rugosum are more sensitive to the treatment than cells of Catharanthus roseus. Other methods for the permeabilization of plant cells should be explored.

<u>Permeabilization For Product Release.</u> Treatment of C. roseus with 5 % DMSO for 30 min is sufficient to make both the plasma membrane and the tonoplast permeable [8,40]. Furthermore, the cells remained viable after this treatment [8]. A process involving intermittant permeabilization for product release was tested in model studies. Indeed, a cyclic process with intermittant product release was possible for alginate and agarose-entrapped cells. An increased yield of product was observed for each cycle due to an increase in biomass during the experiment. The DMSO-treatment of the immobilized cells released about 90 % of the intra-cellularly stored products as compared to extraction with organic solvents.

The experiments described above indicate that it may be possible to utilize immobilized plant cells in a repeated fashion for the production of secondary products. A process of this kind would enable the reuse of the prohibitively

expensive plant cell biomass and, thereby, making such a process more
economically feasible. It should, however, be pointed out that the technique
probably only is applicable to water-soluble products and to certain plant
species.

Permeabilization For Bioconversions. Permeabilization under mild conditions
offers a possibility to utilize specific enzymes in their natural environment for
bioconversions. Such conversions can of course include substrates and/or
coenzymes that cannot penetrate an intact plasma membrane.

The enzyme cathaneamine reductase, converting cathenamine to ajmalicine isomers, was studied in immobilized permeabilized cells of C. roseus according to
Figure 16 [19]. This enzyme requires NADPH for the reduction. Very little ajmalicine was produced from cathenamine when no coenzyme was added. However, when
the oxidized coenzyme (NADP$^+$) was added together with isocitrate a relatively
high product formation was observed. NADPH was in this case generated within the
cells by isocitrate dehydrogenase (an enzyme present within the cells in relatively high activities). Obviously, an enzyme of the primary metabolism can be used
as a coenzyme-regenerating and recycling system when an enzyme of the secondary
metabolism is employed for a bioconversion. This may be of great importance for
certain bioconversions where expensive coenzymes (i.e. NADPH) are required.

Permeabilization of alginate-entrapped cells of M. pruriens by treatment with
10 % (v/v) isopropanol resulted in a three-fold increase in transfomation of
tyrosin to L-DOPA in a batch procedure [58]. However, when the treated cell
preparation was used in a second batch very little hydroxylation activity
remained. This was assumed to be due to loss of enzyme due to washing.

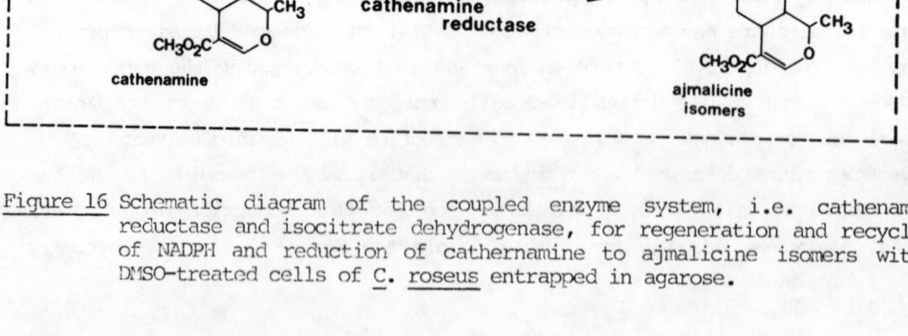

Figure 16 Schematic diagram of the coupled enzyme system, i.e. cathenamine
reductase and isocitrate dehydrogenase, for regeneration and recycling
of NADPH and reduction of cathenamine to ajmalicine isomers within
DMSO-treated cells of C. roseus entrapped in agarose.

REACTORS FOR IMMOBILIZED PLANT CELLS

It can be expected that immobilized plant cells may be utilized in relatively uncomplicated reactors in contrast to freely suspended cells which in most cases require sophisticated reactor design due to their low stress tolerance. Relatively little research has, however, so far been carried out on the design of reactors for immobilized plant cell preparations. Most experiments have been performed in simple batch configurations (i.e. shaker flasks) but there are also examples of packed bed reactors [13,15,25], bubble column reactors [42,56] and a fluidized bed reactor [44].

Recently, membrane reactors were introduced as potential immobilized plant cell reactors [35,53]. In this type of reactor the cells are maintained on one side of the membrane and medium is circulated on the other side of the membrane. This appears to be a convinient way to immobilize plant cells and simultaneoulsy solve the problem of reactor design.

CONCLUDING REMARKS AND PERSPECTIVES

The isolation of high-producing stable cell lines is probably the most important step in the development of a process based on plant cell culture for the production of biochemicals. There are various selection methods available but further developments are required in this area. In particular, methods for the screening of a large number of cultures must be developed. Here cell sorting on a single cell level should be considered.

Genetic transformations of plant cells and somatic hybridazation of plant protoplasts for the improved productivity should also be evaluated. During recent years a number of vectors for the transformation of plant cells have been constructed. These should prove valuable in the future design of cultivated plant cells for the production of target substances.

Secondary metabolism and its regulation in cultured plant cells is another area of future research. A better understanding of plant cell metabolism will lead to possible manipulations of cells to produce enhanced quantities of desired products. Here the use of elicitors on plant cell cultures may be a valuable instrument and such systems should be thoroughly studied in the future.

^{31}P NMR has been shown to be a powerful tool to study phosphate uptake, storage and metabolism in freely suspended and immobilized plant cells. Further studies in this direction should be carried out and the possibility of studying other nuclei should be investigated (e.g. $^{23}Na^+$, $^{39}K^+$ and $^{15}NH_4^+$).

The first report on the immobilization of plant cells appeared only a few years ago [13] and since then an increasing number of reports have occured as summarized above. It is believed that by immobilizing plant cells some problems

194

in utilizing these type of cells on a large scale may be reduced or overcome.

The immobilization allows growth of cells to a certain extent but the production of compounds which are only synthesized during growth would not be advisable with entrapped cells. It is therefore likely that immobilized plant cells only find applications for products that are non-growth associated in a two-stage process (c.f. Figure 12). The biomass produced in suspension cultures is immobilized and used in a second stage under optimal production conditions. The extended stationary phase observed for immobilized plant cells may circumvent some of the problems resulting from the apparent slow growth of cells (biomass generation).

The increased productivity observed in a number of immobilized plant cell systems is not sufficient to overcome the inherent low yield of products in culture. It may, however, be of great importance if high-producing cell lines can be induced to synthesize even more product by immobilization. So far no such investigation has been made.

Cell aggregation occurs in culture and this leads to a wide distribution in aggregate size resulting in problems with agitation of and mass transfer to the cells. These types of problems may be reduced by immobilization since "aggregation" becomes a design parameter. A relatively homogeneous preparation of beads containing the cell aggregates is readily made.

Plant cell culture often show significant changes upon long-term serial passages. For instance, cell lines selected for high productivity of a particular compound frequently lose this capability. The extended stationary phase observed with immobilized cells may to a limited extent also reduce this problem since the cells are utilized in a non-dividing state which should reduce the appearance of genetic changes.

The problem of low shear tolerance of plant cells is eliminated by the entrapment of the cells in a protective polymeric matrix. Mechanical stabilization of cells may result in a simpler reactor design.

The possibility to release intracellularly stored products by intermittant permeabilization of the immobilized cells is still another advantage that may prove valuable in the future development of plant tissue culture technology. This may allow the re-utilization of the biomass over an extended period of time.

The results so far obtained are promising and immobilization of plant cells may prove a valuable tool to overcome or reduce some of the problems encountered in the utilization of plant tissue cultures for the production of complex natural products.

ACKNOWLEDGEMENT

This work was supported in part by the Swiss Natural Science Foundation (3.563-0.83)

195

REFERENCES

1. Alfermann, A.W., Schuller, I. and Reinhard, E. (1980) Planta Med. **40**:218-223
2. Alfermann, A.W., Bergmann, W., Figur, C., Helmbold, U., Schwantag, D., Schuller, I., and Reinhard, E. (1983) In: Plant Biotechnology (Society for Experimental Biology seminar series: 18) pp. 67-74, University Press, Cambridge
3. Bornman, C.H. and Zachrisson, A. (1982) Plant Cell Rep. **1**:151-153
4. Brodelius, P. (1985) In: Applications of isolated enzymes and immobilized cells to biotechnology (A.I. Laskin, ed.) pp. 109-148, Addison-Wesley Publishing Co., Reading, Massachusetts
5. Brodelius, P. (1985) Trends in Biotechnology **3**:280-285
6. Brodelius, P. (1986) Submitted
7. Brodelius, P. and Nilsson, K. (1980) FEBS Lett. **122**:312-316
8. Brodelius, P. and Nilsson, K. (1983) Eur. J. Appl. Microbiol. Biotechnol. **17**:275-280
9. Brodelius, P. and Vandamme, E.J. (1986) In: Biotechnology, Vol. 7 (J.F. Kennedy, ed.) Verlag Chemie Weinheim, in press
10. Brodelius, P. and Vogel, H.J. (1984) Ann. N.Y. Acad. Sci. **434**:496-500
11. Brodelius, P. and Vogel, H.J. (1985) J. Biol. Chem. **260**:3556-3560
12. Brodelius, P., Constabel, F. and Kurz, W.G.W. (1982) In: Enzyme Engineering, Vol. 6 (I. Chibata, S. Fukui and L.B. Wingard, Jr., eds.) pp. 203-204, Plenum Press, New York, N.Y.
13. Brodelius, P., Deus, B., Mosbach, K. and Zenk, M.H. (1979) FEBS Lett. **103**:93-97
14. Brodelius, P., Deus, B., Bosbach, K. and Zek, M.H. (1980) In: Enzyme Engineering, Vol. 5 (H.H. Weetall and G.P. Royer, eds.) pp. 373-381, Plenum Press, New York, N.Y.
15. Brodelius, P., Deus, B., Mosbach, K. and Zenk, M.H. (1981) European Patent Application 80850105.0
16. Cabral, J.M.S., Cadete, M.M., Novais, J.M. and Cardoso, J.P. (1983) Poster 1:3 presented at 7th International Conference on Enzyme Engineering, September 25-30, 1983, White Haven, PA.
17. Ebel, J., Schmidt, W.E. and Loyal, R. (1984) Arch. Biochem. Biophys. **232**:240-248
18. Eilert, U., Ehmke, A. and Wolters, B. (1982) Planta Med. **45**:155
19. Felix, H., Brodelius, P. and Mosbach, K. (1981) Anal. Biochem **116**:462-470
20. Freeman, A. and Aharonowitz, Y. (1981) Biotechnol. Bioeng. **23**:2747-2759
21. Fujita, Y., Hara, Y., Suga, C. and Morimoto, T. (1981) Plant Cell Rep. **1**:61-63
22. Fujita, Y., Tabata, M., Nishi, A. and Yamada, Y. (1982) In: Plant Tissue Culture 1982 (a: Fujiwara, ed.) pp. 399-400, Japanese Association for Plant Tissue Culture, Tokyo
23. Funk, C. and Brodelius, P. (1985) unpublished
24. Funk, C., Gügler, K. and Brodelius, P. (1986) Phytochemistry, in press
25. Furuya, T., Yoshikawa, T. and Taira, M. (1984) Phytochem. **23**:999-1001
26. Gadian, D.G. and Radda, G. (1981) Annu. Rev. Biochem. **50**:69-85
27. Galun, E., Aviv, D., Dantes, A. and Freeman, A. (1983) Planta Med. **49**:9-13
28. Gustine, D.L., Sherwood, R.T. and Vance, C.P. (1978) Pland Physiol. **61**:226-231
29. Haldimann, D. and Brodelius, P. (1986) Submitted
30. Heinstein, P.F. (1985) J. Nat. Prod. **48**:1-9
31. Hulst, A.C., Tramper, J., Brodelius, P., Eijkenboom, L.J.C. and Luyben, K.Ch.A.M. (1985) J. Chem. Tech. Biotech. **35B**:198-204
32. Jirku, V., Macek, T., Vanek, T., Krumphanzl, V. and Kubanek, V. (1981) Biotechnol. Lett. **3**:447-450
33. Jones, A. and Veliky, I.A. (1981) Eur. J. Appl. Microbiol. Biotechnol. **13**:84-89
34. Jones, A. and Veliky, I.A. (1981) Can. J. Bot. **59**:2095-2101
35. Jose, W., Pedersen, H. and Chin, C.K. (1984) Ann. N.Y. Acad. Sci. **413**:409-412
36. Klein, R.M. (1960) Econ. Bot. **14**:286-289

37. Lambe, C.A. and Rosevear, A. (1983) Proceedings of Biotech83, London, May 4-6, 1983, pp. 565-576
38. Lindsey, K., Yeoman, M.M., Black, G.M. and Mavituna, F. (1983) FEBS Lett. 155:143-149
39. Linsefors, L. and Brodelius, P. (1983) unpublished
40. Lundberg, P., Linsefors, L., Vogel, H.J. and Brodelius, P. (1986) Plant Cell Reports 5:13-16
41. Martin, J.B., Bligny, R., Rebeille, F., Douce, R., Leguay, J.-J., Mathieu, Y. and Guern, J. (1982) Plant Physiol. 70:1156-1161
42. Moritz, S., Schuller, I., Figur, C., Alfermann, A.W. and Reinhard, E. (1982) In: Proc. 5th Intl. Cong. Plant Tissue and Cell Culture (A. Fujiwara, es.) pp. 401-402, Japanese Association for Plant Tissue Culture, Tokyo
43. Morris, P. and Fowler, M.W. (1981) Plant Cell Tiss. Org. Cult. 1:15-24
44. Morris, P., Smart, N.J. and Fowler, M.W. (1983) Plant Cell Tissue Organ Culture, 2:207-216
45. Nakajima, H., Sonomoto, K., Usui, N., Sato, F., Yamada, Y., Tanaka, A. and Fukui, S. (1985) J. Biotechnol. 2:107-117
46. Neumann, K.-H., Barz, W. and Reinhard, E., eds. (1985) Primary and Secondary metabolism of plant cell cultures, Springer-Verlag, Berlin, Heidelberg
47. Nilsson, K., Birnbaum, S., Flygare, S., Linse, L., Schroder, U., Jeppsson, U., Larsson, P.-O., Mosbach, K. and Brodelius, P. (1983) Eur. J. Appl. Microbial. Biotechnol. 17:319-326
48. Reinhard, E. and Alfermann, A.W. (1980) In: Advances in Biochmical Engineering, Vol 16 (A. Fiechter, ed) pp. 49-83, Springer-Verlag, Berlin and Heidelberg
49. Roberts, J.K.M., Wemmer, D., Ray, P.M. and Jardetsky, O. (1982) Plant Physiol. 69:1344-1347
50. Rokem, J.S., Tal, B. and Goldberg, I. (1985) J. Nat. Prod. 48:210-222
51. Rosevear, A. (1981) European Patent Application 81304001.1
52. Roesevear, A. and Lambe, C.A. (1982) European Patent Application 82301571.4
53. Shuler, M. (1981) Ann. N.Y. Acad. Sci. 369:65-79
54. Shuler, M. and Hallsby, A.G. (1983) Paper #76b at AIChE 1983 Summer National Meeting, Denver, Colorado, August 28-31
55. Staba, E.J. ed (1980) Plant tissue culture as a source of biochemicals, CRC Press, Boca Raton, Fl.
56. Veliky, I.A. and Jones, A. (1981) Biotechnol. Lett. 3:551-554
57. Vogel, H.J. and Brodelius, P. (1984) J. Biotechnol. 1:159-170
58. Wichers, H.J. (1985) PhD Thesis, Rijkuniversiteit te Grodningen, Groningen, The Netherlands
59. Wijnsma, R., Go, J.T.K.A., van Weerden, I.N., Harkes, P.A.A., Werpoorte, R. and Baerheim-Svendsen, A. (1985) Plant Cell Rep. 4:241-244
60. Wolters, B. and Eilert, U. (1983) Deut. Apo. Zeit. 123:659-667
61. Wray, V., Schiel, D. and Berlin, J. (1983) Z. Pflanzenphys. 112:215-220
62. Yamamoto, Y., Mizuguchi, R. and Yamada, Y. (1982) Theor. Appl. Genet. 61:113-116

THE APPLICATION OF CONTINUOUS THREE PHASE FLUIDIZED BED BIOREACTORS TO THE PRODUCTION OF PHARMACEUTICALS.

Leo A. Behie* and G. Maurice Gaucher+
*Department of Chemical Engineering, +Division of Biochemistry
The University of Calgary
Calgary, Alberta, Canada
T2N 1N4

ABSTRACT

Classic three phase fluidized bed bioreactors have proven to be an excellent reactor type for the continuous production of pharmaceuticals by immobilized cells. Products studied in our laboratories include two antibiotics, patulin and penicillin-G, which required widely different cell specific growth rates to achieve high productivities. For these antibiotics two immobilization protocols were developed using two support matrices (carrageenan, a polysaccharide gel and Celite beads, porous diatomaceous earth) with the important operational result that the cells remained anchored to the matrix and the medium essentially free of cells. Even at high cell densities (>30 g/L), no oxygen mass transfer limitations were observed. Finally, some results for the batch production of monoclonal antibodies (MAb) from immobilized hybridoma cells are presented. Using agarose as the support matrix, maximum MAb concentrations were 250% of concentrations achieved for free cell suspension cultures in spinner flasks. However, the percent viability of the immobilized hybridoma cells after 100 h was very low.

KEYWORDS

Bioreactor, fluidized bed, immobilized cells, antibiotics, monoclonal antibiodies, continuous.

SCOPE

The development of three phase fluidized bed bioreactors is expected to have an enormous impact on the pharmaceutical industry. This is particularly true for antibiotics which are among the most important products produced by biological processes. For example, the current world production of Penicillin-G exceeds 20,000 tonne/year and bioreactor vessels up to 100,000 L are not uncommon. Despite the large scale of production, batch and fed-batch submerged free cell fermentations remain the

197

predominant modes of operation in industry.

The main problem of free cell batch of fed-batch antibiotic fermentations is the severe limitations in oxygen transfer (from gas bubbles to microbes dispersed in the liquid medium) since, as the amount of biomass (i.e. free mycellia or flocs) in the reactor increases, the liquid phase becomes very viscous and difficult to agitate. On the other hand, in a continuous operation when the growth rate of cells is less than their removal rate, the cells are washed out of the reactor and antibiotic production gradually ceases. Unless these problems are overcome, no economic benefit can be expected by going to large scale continuous operation.

Immobilized cell technology provides a simple and promising alternative to free cell fermentations, the principle differences being superior mass transfer at high cell densities and the capability to run the fermentation in a continuous fashion. Recently, our biotechnology group at the University of Calgary has successfully developed a three phase fluidized bed bioreactor where the limitations of oxygen transfer are completely alleviated (Berk et al., 1984a, 1984b, 1984c; Jones et al., 1984, 1986). The growing cells containing the necessary biocatalysts (enzymes) are immobiized inside inert support matrices. This makes up the solid phase. With concurrent air, the fluidized bed can be described as having bubble supported solids in which the liquid (medium) flow is far below its minimum fluidization velocity.

Apart from new strain development, the proper selection of the feeding medium and state-of-the-art computer control of the fermentation process are the only means to achieve high production yields which could make the continuous operation an economic success. This paper presents data on mass transfer effects in fluidized bed fermentors, specific rates of reaction for immobilized cell penicillin-G and patulin production, and a mathematical model used to evaluate the performance of the new process under computer control. Finally, some results are given for the very different process of monoclonal antibody (MAb) production by mammalian cells immobilized in agarose.

CONCLUSIONS AND SIGNIFICANCE

An immobilized cell fluidized bed bioreactor for antibiotic production has been run continuously for over 10 days using two different support matrices. With our control system in place, the productivity of the bioreactor can be increased substantially to make continuous fermentations economically attractive. With immobilized mammalian cells run in a batch reactor, severe problems with cell viability beyond 100 h were observed although antibody concentrations achieved were much higher than those observed in spinner flasks using free cell suspension cultures.

INTRODUCTION

The advent of immobilized whole cell technology has generated considerable interest in the field of biotechnology and several applications have been extrapolated from the laboratory to the pilot plant or industrial scale (Chibata and Tosa, 1981). A number of advantages are inherent in the use of immobilized cells. These include the increased lifetime of the biocatalyst in the immobilized state, increased productivity, and from a biochemical engineering point of view, a means of overcoming the washout of continuous cultures and the mass transfer problems usually associated with growth at high cell densities. This change in catalyst conformation from either single or mycelial cells to that of larger spherical or irregularly shaped particles has also led to the development of specialized bioreactors, for example bubble columns, fluidized beds or airlift reactors to accomodate these biocatalysts.

Fungi are an important source of a variety of industrially important products e.g. antibiotics, steroids and enzymes. However, an unfortunate feature of the growth of these mycelial organisms is the high viscosity of the culture broth in conventional submerged cultures. This results in a decrease in the mass transfer capabilities of conventional stirred tank reactors for oxygen and to a lesser degree for nutrients such as carbon and nitrogen sources. To overcome these problems the agitation and aeration rates can be increased but this often results in severe shear stresses being imposed on these filamentous cells. Consequently, cell viability decreases and thus productivity declines.

An alternative to mycelial growth is that of pelleted growth. The latter can be induced by manipulation of the growth medium or environmental factors such as pH (Kim et al., 1982; Konig et al., 1982; Wittler et al., 1983). Alternatively, the rheology of the broth can be similarly changed by using immobilized cells, which can be regarded as mycelial pellets since growth is often restricted to the external surface of the beads. Thus, immobilized fungal cells can serve as a convenient method of directed pellet formation which does not suffer from the often unpredictable media or environmental manipulations necessary for induced pelletization in normal cultures.

Our studies have been directed to exploring the potential uses of immobilized cells for the production of pharmaceuticals in three phase fluidized beds and to an understanding of the engineering aspects of bioreactor design for the use of such biocatalysts. As model systems we have examined the production of the mycotoxin, patulin by Penicillium urticae (Berk et al., 1984b, 1984c) and the industrially important antibiotic, penicillin-G by Penicillium chrysogenum (Berk et al., 1984a, Jones et al., 1986). Several aspects of these fungal cells immobilized in carrageenan have been examined and a medium for the semi-continuous or continuous production of penicillin-G over a period of 10-15 days has been developed. As an extension of these studies, attention has now been turned to the use of Celite beads as an immobilization matrix (Jones et al., 1986). Celite, which is available commercially in a variety of particle sizes, is also used as a support in hetrogeneous catalysis systems. Its use as a support matrix for Penicillium chrysogenum was described by Gbowenyo and Wang (1983a, 1983b). These authors showed that the cells would successfully produce penicillin in a tower loop reactor. Of interest was their finding that normal cultures had a much reduced capacity for antibiotic production under similar conditions of growth.

This paper presents data on two similar types of cell support matrices used for antibiotic production in a 1.2 L three phase fluidized bed bioreactor running continuously. Furthermore, a model of the system is presented and is subsequently used to simulate the performance of the bioreactor for penicillin-G production when under computer control. Finally, some results are given for the quite different process of monoclonal antibody (MAb) production by mammalian cells immobilized in agarose.

EXPERIMENTAL MATERIALS AND METHODS

I Antibiotics - Patulin and Penicillin-G.

Immobilization in Carrageenan. Two cell support matrices were used for patulin and penicillin-G production. Complete protocols are described by Gaucher and Behie (1986) and Jones et al. (1986). The first, κ-carrageenan, is an inert polysaccharide derived from seaweed and sold commercially by the FMC Corporation. It had many attributes as a support. Not only were the 4% carrageenan beads very robust but they allowed the fungi to grow inside the beads without affecting the morphology of the cells. Furthermore, the spherical beads, having a mean diameter of either 3.2, 3.0, 2.7 or 1.8 mm, had a very narrow size distribution (standard deviation 0.15 mm). The carrageenan beads were cast using a device developed specifically for this study (Berk et al., 1984b; Gaucher and Behie, 1986). About 20,000 beads were used in the 1.2 L reactor. The culture of Penicillium chrysogenum E15 (ATCC 26818) producing penicillin-G was obtained as freeze-dried ampoules from Dr. S.W. Queener of the Eli Lilly Co. Ltd., while Peniccillium urticae (NRRL 2159A) was used for patulin production.

Immobilization in Celite Beads. Celite is basically diatomaceous earth which is sold in different mesh sizes and different pore size distributions. In this study, Celite R-630 (Johns Manville Co. Ltd.), was used with an initial mean diameter of 410 μm, pore diameter of 8.0 μm, specific pore volume 1.70 cm^3/g and bulk density of 0.27 g/cm^3. All preparation details can be found in Jones et al., (1986). Only penicillin-G production was examined in this part of the study.

Three Phase Fluidized Bed Reactor. The reactor had a diameter of 7.5 cm and a working volume of 1.2 L. At the reactor top, the diameter increased to provide a solids disengaging section. Saturated air ws introduced uniformly at the bottom through a high pressure drop porous distributor at a superficial velocity of 1.4 cm/s. The reactor was enclosed in a constant temperature jacket that maintained the temperature at 26 ° C and 28 ° C for penicillin and patulin production respectively.

The liquid was fed through a tube just above the distributor. In the production medium for penicillin production all necessary nutrients were present, including: KH_2PO_4 (6.8 g/L), K_2SO_4 (3.95 g/L), phenylacetic acid (0.3 g/L) and finally, NH_4Cl (22 mM) and lactose (10 g/L) as nitrogen and carbon source respectively. All microbes were confined in the beads at all times with the medium essentially free of cells. Further, the medium pH was kept in the range of 6.2 to 6.8 in order to minimize penicillin-G hydrolysis to 6-APA (Pirt, 1985). Finally, the mean residence time of the medium in the reactor was in the range 24 to 50 h. More details on the design and operation of the reactor as well as all measurement techniques can be found in Berk et al. (1984b, 1984c). The production medium for patulin production can be found in Berk et al. (1984b).

II Monoclonal Antibodies (MAb).

Mouse hybridoma cells producing monoclonal antibody to human transferrin (anti-HT cells) were received from Dr. T. Pearson, Department of Microbiology and Biochemistry, University of Victoria, Victoria, B.C., Canada. Mouse hybridoma cells producing monoclonal antibody to the B antigen of human erythrocytes were received from

Chembiomed Ltd.

All growth medium components were as described previously (Adamson et al., 1983, 1986a).

Maintenance of Cells. Hybridoma cells producing monoclonal antibody to human transferrin (anti-HT cells) and group B antigen of human erythrocytes (anti-B cells) were maintained in T-flask cultures in RPMI-1640 growth medium supplemented with 10% fetal calf serum (FCS), 2 mM glutamine, 100 IU penicillin-G/mL and 100 μg streptomycin sulfate/mL in a humidified atmosphere of 5% CO_2 in air at 37 °C.

Immobilization of Hybridoma Cells.

For immobilization of hybridoma cells exponentially growing cells (3-5 x 10^5 cells/mL) were harvested from growth medium by centrifugation at 200X g. The cells were resuspended in 15 mL of RPMI 1640 containing 20% FCS, 4 mM glutamine, 100 IU penicillin-G/mL and 100 μg streptomycin sulfate/mL. This cell suspension was mixed thoroughly with an equal volume of agarose solution (2% w/v; phosphate buffered) and cast into beads of 2.5 mm equivalent diameter. The beads were suspended in an equal volume of growth medium (25 mL) in a spinner flask and stirred at 50 rpm in a humidified atmosphere of 5% CO_2 in air at 37 °C.

Viability Estimates.

Viability was usually estimated by trypan blue dye exclusion. Slices of agarose were cut from beads using a razor, or small columns were bored out of the beads using a 100 μL microsampling pipette. Sectons were placed in approximately 0.5 mL of trypan blue for 5 min at 37 °C to allow the dye to penetrate the agarose. In experiments where staining was not carried out, viability was estimated by microscopic examination. Viability of cells was judged by the degree of granularity. Good agreement between stained and unstained samples was routinely observed.

Nutrient Maintenance Systems. Dialysis experiments involving immobilized cells employed the same apparatus as previously described (Adamson, 1983). Perfusion experiments were carried out in a modified Bellco spinner flask with an overflow outlet. Medium was perfused as required by use of an LKB perstaltic pump from a medium reservoir kept in a humidified atmosphere of 5% CO_2 in air at 37 °C.

Antibody and DNA Assays.

Anti-HT and anti-B antibody titres of culture supernatants were determined by enzyme linked immunosorbent assay (ELISA) as previously described (Adamson, 1983), excecpt that BSA coupled antigen was used in the anti-B assay (Bundle et al., 1982).

The DNA content within the gel matrix was determined after harvesting the cells. A sample of gel (approximately 0.2 g) was partially dried by blotting on a Kimwipe, weighed accurately and transferred to a 15 mL Corning polystyrene centrifuge tube. The beads were suspended in 12 mL of phosphate buffered saline (PBS) and the tubes placed in a boiling water bath for exactly 2 min (the contents were mixed after 1 min).

The cells were collected on Whatman Glass Microfibre Filters (GF-A) and the filters washed three times with 15 mL of preheated (50 °C) PBS. The filters were suspended in 5 mL of 50 mM sodium phosphate buffer, pH 7.4, Containing 2M NaCl in round bottom glass centrifuge tubes (2.25 cm inside diameter). The samples were sonicated on ice using a Branson sonifier at the output setting 4 on a 50% duty cycle for 12 seconds. After a 30 s period for cooling a further 12 s sonication was carried out. The fragmented filter material was removed by centrifugation and 2 mL samples were taken from the clear supernatant for assay. The DNA content of the gel was determined by the Hoechst-33258 DNA binding assay with reference to a mouse DNA standard. Recoveries of 80% were attained using this procedure, as determined by comparison with free cells which were disrupted and assayed in the same manner. All data have been corrected for this level of recovery.

The cellular DNA content was determined as follows: aliquots of 1.5 mL were removed from cultures of exponentially growing anti-HT hybridoma cells and centrifuged in an Eppendorf microcentrifuge (model 5412) for 3 min. The cell pellet was resuspended in 5 mL of 50 mM sodium phosphate buffer, pH 7.4, containing 2 M NaCl and sonicated as described above. Cell number analysis was determined by counting the original culture on a hemocytometer.

Metabolite Assays.

Glucose and lactate in hybridoma culture media were measured using kits supplied by Worthington and Sigma respectively, essentially as described in the manufacturer's instruction. In both cases recommended volumes were scaled down to 1 mL.

Quantities of amino acids in aliquots of culture supernatant were determined by automatic amino acid analysis after deproteinization. To 2 mL of culture medium was added 100 μL of 5.3 mM norleucine, followed by 5 mL of cold 95% ethanol (67% final). The samples were stored overnight at 5°C and the following day the resulting precipitate was removed by centrifugation at 12 000 x g at 5 °C. Ethanol was removed by evaporation and 5 mL of deionized wter was added back. The samples were lyophilized and the resulting powder taken up in the first buffer used in amino acid analysis. The original concentration of amino acids was calculated relative to the norleucine internal standard. Glutamine was determined as glutamic acid after hydrolysis with 3 M HCl at 110 °C for 2 h. The role of amino acids in the metabolism of hybridoma cells is discussed by Adamson et al. 1986a.

Dissolved Oxygen Measurement.

The relative dissolved oxygen concentration in hybridoma culture medium was measured with a New Brunswick Scientific D.O. Analyser. The instrument was calibrated to 100% with culture medium which had been fully equilibrated in a 5% CO_2/95% air atmosphere and to 0% with culture medium, in the absence of serum, which had been flushed for 5-10 min with nitrogen.

EXPERIMENTAL RESULTS AND DISCUSSION

I Antibiotics - Patulin and Pencillin-G.
Carrageenan Beads.

   In the continuous runs presented here, three different levels of lactose concentration
in the feed (10, 5 and 2 g/L) were used. The penicillin concentration in the reactor
effluent versus time, at a liquid residence time of 24 h, is shown in Figure 1 for a bead

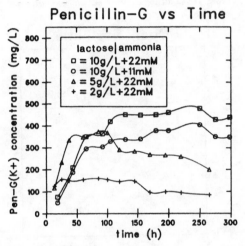

Figure 1. Concentration of Penicillin-G(K^+) in the Fluidized Bed Effluent for Various
Liquid Mediums (3.0 mm Carrageenan beads)

size of 3.0 mm. Also shown is the effect of reducing the nitrogen source in the feed
stream from 22 mM to 11 mM (NH_4Cl). The results show that when the lactose
concentration in the feed is 2 g/L, the penicillin concentration in the effluent stays
constant at 150 mg/L by 300 h. On the other hand, a rapid increase in the production
of penicillin is observed during the first 40 to 60 h when the higher feed concentrations
of lactose (10 and 5 g/L) are used. Decreasing the concentration of the ammonium ion
from 22 to 11 mM results in a decrease in the penicillin concentration.

   The effluent concentrations of ammonia and lactose were also measured (Kalogerakis
et al., 1986a). When the lactose content in the feed was low (5 or 2 g/L), its residual
concentration in the effluent was zero at all times indicating that it was the limiting
substrate. This was never the case with the high lactose concentration feeds. On the
other hand, there was a nitrogen limitation in the reactor when the ammonium ion was
reduced to 11 mM.

   An unfortunate characteristic of P. chrysogenum for maintaining its capability to
synthesize penicillin is that it requires the formation of new cells. As expected, higher
lactose concentrations in the feed resulted in a higher biomass increase in the reactor.
The highest biomass concentration was obtained when neither lactose nor ammonium

were in limiting concentrations. It is interesting to note that the highest penicillin concentration was also obtained for the same conditions in the feed. As the biomass was reduced due to nutrient limitations, the penicillin production was also seen to decrease.

In Figures 2 and 3 the specific reaction rates (q_p) and yield coefficients (Y) for penicillin are shown versus time for two different bead sizes (3.0 and 1.8 mm). The curves were generated from a material balance on the reactor taking into account the accumulation terms. Note that, although the reactor was running continuously, it never achieved steady-state conditions.

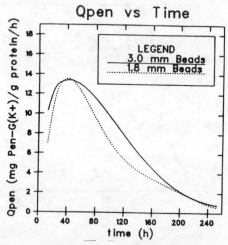

Figure 2. Specific Rates of Reaction for Penicillin-G Production Using Two Carrageenan Bead Sizes

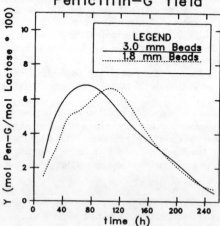

Figure 3. Variation of Penicillin-G Yield Coefficient (Y) for Two Carrageenan Bead Sizes

The dominant mass transfer resistance in the reactor was found to be diffusion within the solid support matrix. Mass diffusivities in carrageenan have been measured and confirm this conclusion (Berk, 1984b; Kalogerakis et al., 1986c). However, most of the cell growth occurs on the surface of the beads (final layer thickness was less than 100 microns) during the penicillin production phase even though spores are initially distributed uniformly throughout the beads. Thus, by going from 3.0 mm beads to 1.8 mm, the reaction rates and yields remained the same as shown in Figures 2 and 3. Hence, the bead mass transfer resistance was not significant. In other words, the reaction rates were very slow and the reactor was under kinetic control. Although cell spores were initially uniformly dispersed throughout the bead, only the growing cells on the surface produced penicillin.

Figure 4 shows that for patulin this is not the case. The unsteady-state material balance yields the specific global reaction rate as a function of time. Much higher rates are observed for the 2.7 mm

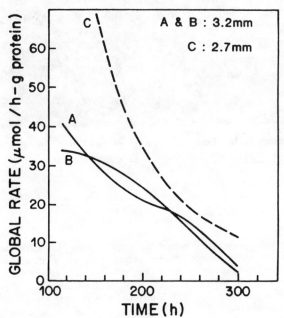

Figure 4. Effect of Bead Diameter on the Decline in Patulin Production During Continuous Reactor Runs

beads compared to the 3.2 mm beads indicating more severe mass transfer limitations for the larger beads. The basic reason that the patulin reactor was so different than the penicillin reactor can be attributed to the low cell growth rate. In fact, once the production phase is reached for patulin production, the cell growth is practically turned off by nitrogen starvation (i.e. cell specific growth rate, $\mu = 0.0007$ h[-1]). More details can be found in Berk et al., 1984c. This means that significant amounts of patulin are produced within the beads and can only diffuse out with difficulty. The intra-particle mass transfer resistance is significant. On the other hand, for high penicillin-G production rates, cells must grow at a much higher rate (i.e. cell specific growth rate, μ = 0.014 h^{-1}), an important problem discussed by Pirt (1985). As a result, for penicillin production, most of the cell growth occurs on the surface of the carrageenan beads. This is clearly shown in Figure 5 which is a transmission light micrograph of a 10 micron thick slice of a carrageenan bead taken at 163 hours into the run.

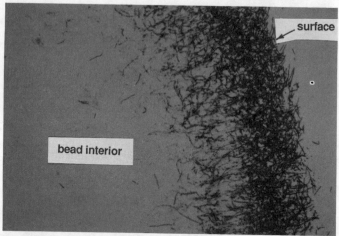

Figure 5. Transmission Light Micrograph of a 10 Micron Slice Through the Center of a 3.0 mm Carrageenan Bead Sampled at 160 h (P. chrysogenum)

Celite Beads.

In a recent penicillin-G study, Jones et al., (1986) found that Celite beads were five times more productive on a reactor volume basis. This result is not surprising since Celite beads were able to support much higher cell densities (more than 30 g/L), and hence, they perform much better in a three phase fluidized bed. Nevertheless, the maximum specific reaction rates (q_p^{max} observed for both carrageenan and Celite were practically the same at about 13.5 mg Pen-G(K$^+$)/g protein/h, a value comparable to that found in submerged free cell fed-batch fermentations for this E15 generation of microbe.

Figure 6 shows a microscope photograph of the Celite beads used in the Jones (1986) study. On the left initial Celite R-630 particles are shown (-30+50 mesh cut; mean diameter = 410 micron). On the right of the photograph are shown a number of beads sampled at 160 h of a continuous penicillin run showing clearly the increased size due to cell growth. Note that each of the beads shown on the right move independently in the reactor and not as clumps.

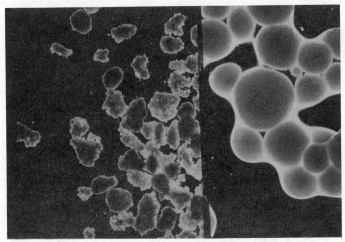

Figure 6.   Photographs - Left: Celite R-630 Beads Without Cells
(mean diameter = 410 micron)
Right: Sample of Celite Beads Taken at
160 h of a Continuous Pen-G Run
(P. chrysogenum)

Increased Productivity Under Computer Control.

Recently, we designed a simple noninteractive control system which allows the user to control the cell specific growth rate and the penicillin concentration in the effluent stream, to any desired values by manipulating the feed flowrate and the substrate concentration in the feed (Kalogerakis et al., 1986b).

The only limitation of the system is that, as the cells in the reactor grow continuously at the specified growth rate, there will come a certain point in time where the bed will defluidize. The way around this problem is to remove beads and add new ones in the reactor continuously or in an intermittent fashion so that the average cell concentration in the reactor remains constant. This will allow the fermentations to run for extended periods of time until other factors come into play (eg. contamination). The controller has two other feedback loops and thus posesses self correcting capabilities to modelling errors (which can be viewed as slowly varying model parameters).

The cell concentration in the reactor can easily be kept constant by introducing an independent P+I controller which will manipulate the bead removal rate. Such a modification to the reactor will allow the system to reach steady-state conditions where the average cell concentration, specific growth rate and penicillin concentration remain constant at the values selected by the process engineer.

The underlying idea here is to select the desired specific cell growth rate (0.010 to 0.015 h^{-1}) where the penicillin production rate is maximum and at the same time keep the average cell concentration at a constant but high value (35 to 50 g/L) so that the penicillin productivity ($q_p.x$) is maintained very high at all times during the fermentation. Furthermore, the penicillin concentration in the effluent will be kept constant at the specified concentration level by the control system so that the potency of the product stream is high. The actual value will be the result of experimental investigations so that any possible inhibition effects or side reactions are avoided. In summary, the system will be maintained under tightly controlled environmental conditions which will correspond to the best operating conditions for penicillin production. It is assumed here that pH and temperature will also be kept constant at optimally selected values by conventional PID controllers. Dissolved oxygen is not a problem for our reactor since we work with high air flows and the broth remains virtually free of cells at all times.

The dynamic behavior of the fermentation process is modelled by considering unsteady state component mass balances (Kalogerakis et al., 1986b), namely

$$dx/dt = (\mu - F)x \qquad (1)$$
$$ds/dt = \sigma x + D(s_f - s) \qquad (2)$$
$$dp/dt = q_p x - Dp \qquad (3)$$
$$dz/dt = -bq_p x + D(z_f - z) \qquad (4)$$

where x,s,p and z are the cell, limiting substrate (glucose), penicillin and precursor (PAA) concentrations (g/L) in the reactor. D is the dilution factor (volumetric feedrate over liquid volume, h^{-1}, s_f and z_f are the substrate and precursor concentrations (g/L) in the feed stream and F is the bead removal rate expressed as a volumetric cell removal flow over liquid volum (h^{-1}).

The cell specific growth rate, μ, is assumed to follow the usual Contois type kinetics, i.e.,

$$\mu = \mu_{max}s/(K_s x + s) \qquad (5)$$

The specific uptake rate, σ, is related to the growth rate by (Pirt, 1985)

$$\sigma = \mu/Y_G + m + q_p/Y_p \qquad (6)$$

where Y_G is the cell growth yield, m is the maintenance energy and Y_p is the penicillin from substrate yield (carbon and ATP source).

The specific penicillin production rate, q_p, is related to the average cell age, λ, by (Fishman and Biryokov, 1974; Holmberg and Ranta, 1982)

$$q_p = q_p^{max}a \lambda \exp(1 - a \lambda) \qquad (7)$$

where $d\lambda/dt = 1 - \mu\lambda$ (8)

A typical transient response of the system under computer control is shown in Figure 7. As seen, after some time we reach and maintain steady-state conditions where

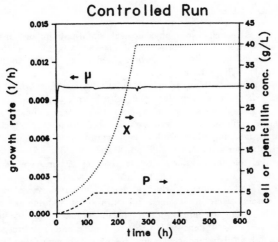

Figure 7. Cell Growth Rate, Penicillin-G Concentration and Cell Concentration Transients for a Typical Fermentation Run Under Computer Control

penicillin is produced at a high rate. The amount of penicillin collected in the exit stream for the above run, as well as for other penicillin concentrations in the effluent, is shown in Figure 8. As can be seen, there is a trade off between penicillin concentration in the exit stream and amount produced.

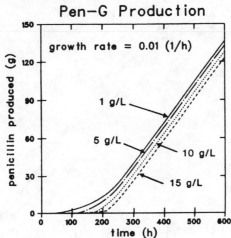

Figure 8. Total Penicillin Produced at Different Penicillin Concentrations in the Effluent Stream - Reactor Under Computer Control

II Monoclonal Antibodies (MAb)

Since there is a paucity of physiological information on mammalian cells, mathematical modelling of such systems is in its infancy (Adamson and Schmidli, 1986b). Perfusion and dialysis culture systems have been used to maintain cells in a nutritionally adequate environment. The ultimate aim in these systems is to establish a high cell density culture in a rich growth medium, which usually contains 5-10% serum; then, on reaching maximum cell density, to perfuse the culture with a production or maintenance medium which is usually low in protein content (0-1% serum or composed of serum substituents). This system allows the continuous production of product from high cell density cultures into medium which is low in contaminating protein, thus significantly enhancing productivity per unit time and purification yields. An evaluation of three commercial media available for hybridoma cellsis given by Adamson et al., (1986a).

Adamson et al., (1983) showed clearly that relatively high levels of nutrients and low levels of waste products were maintained in cultures of hybridoma cells by dialysis against growth medium supplemented with fetal calf serum. This resulted in a greater than ten fold increase in cell density (to 10^7 cell/mL) and a corresponding ten fold increase in monoclonal antibody concentration compared to conventional batch cultures. Because the dialysis tubes used in this study excluded substances greater than a molecular weight of 10 000, the implication was that higher molecular weight substances in serums (proteins) may not be required for cell cultivation.

The data we wish to report on in this paper is shown in Figure 9. With our hybridoma cells immobilized in agarose, we observed a serious problem with cell viability.

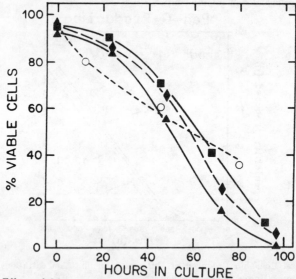

Figure 9. The Effect of Maintenance of Nutrients by Parfusion (at 15 mL/h ■, at 30 mL/h ▲) or Dialysis (◆) on the Viability of Immobilized Anti-B Hybridoma Cells. Batch Culture Results Also Shown (O)

211

Even with perfusion of dialysis systems in place, no improvement was observed compared to a batch system. Replication, as measured by the DNA content within the gel matrix, was physically restricted by immobilization (the doubling time was 40 h versus 16 h in free cell suspension culture). In other words, the cells were physically constrained by the agarose environment. Some limited cell division was observed, however of the cells that did divide (20-30%) they did so only once to form cell doublets.

There was one very positive result for hybridomas immobilized in agarose. When cell densities of about 2×10^6 cells/mL were employed, antibody concentration were 250 percent of free cell suspension cultures. Hence, if the cell viability problem could be overcome, the immobilization of hybridoma in agarose could lead to an important production method.

ACKNOWLEDGEMENTS

This work was supported by the Natural Science and Engineering Reaserch Council of Canada and Chembiomed Ltd.

LITERATURE CITED

Adamson, S.R., S.L. Fitzpatrick, L.A. Behie, G.M. Gaucher and B.H. Lesser, Biotech. Letters, 5 (9), 573 (1983).

Adamson, S.R., L.A. Behie, G.M Gaucher and B.H. Lesser, a chap. in the book Commercial Production of Monoclonal Antibodies, Ed. S. Seaver, Marcel Dekker, New York (in press, 1986a).

Adamson, S.R. and B. Schmidli, Can. J. Chem. Eng. (August, 1986b). Berk, D., L.A. Behie, A. Jones, B.H. Lesser and G.M. Gaucher, Proc. of 34th Can. Chem. Eng. Conf., Quebec City, Canada, 279 (1984a).

Berk, D., L.A. Behie, A. Jones, B.H. Lesser and G.M. Gaucher, Can. J. Chem. Eng., 62, 112(1984b).

Berk, D., L.A. Behie, A. Jones, B.H. Lesser and G.M. Gaucher, Can. J. Chem. Eng., 62, 120 (1984c).

Bundle, D.R., M.A.J. Gidney, N. Kassam and A.F.R. Rahman, J. Immunol., 129, 678 (1982).

Chibata, I. and T. Tosa, Ann. Rev. Biophys. Bioeng., 10, 197 (1981).

Fishman, V.M. and V.V. Biryokov, Biotechn., 16 647 (1974). Gaucher, G.M. and L.A. Behie, a chap. in the book Methods in Enzymology Ed. K. Mosbach, Academic Press Inc.(in press).

Gbewonyo K. and D.I.C. Wang, Biotechn. Bioeng., 25, 967 (1983a). Gbewonyo K. and D.I.C. Wang., Biotechn. Bioeng., 25, 2873 (1983b).

Holmberg, A. and J. Ranta, Automatica, 18, 181(1982). Jones, A., R. Siqueira, D. Berk, T. Razniewska, B.H. Lesser, L. A. Behie and
G.M. Gaucher, Can. J. Microbiol., 30, 475(1984).

Jones, A., D.N. Wood. T., Razniewska, G.M. Gaucher and L.A. Behie, Can. J. Chem. Eng. (August, 1986).

Kalogerakis, N., L.A. Behie and G.M. Gaucher, Fluidization V 611 (1986a),
referred Proc. of Fifth Eng. Found. Conf. on Fluid. held in Elsinore, Denmark (May, 1986).
Kalogerakis, N., T. Linardos, L.A. Behie, W.Y. Svrcek and G.M. Gaucher, Can. J. Chem. Eng. (August, 1986b).

Kalogerakis, N., L.A. Behie, G.M. Gaucher and J. Thibault, accepted for
presentation at the Annual AIChE meeting, Session: "Transport Effects in Bioreactors", Miami Beach, Florida (Nov. 1986c).

Kim, J.H., J.M. Lebeault and M. Reuss, Europ. J. Applied Microb. Biotechnol.
18, 11(1982).

Konig, B., K. Schugerl and C. Seewald, Biotech. Bioeng., 24, 259(1982).

Pirt, S.J., Kem. Ind., 34, 13(1985).

Wittler, R., R. Matthes and K. Schugerl, Europ. J. Applied Microb. Biotech.,
18, 17(1983).

HYDROLYSIS OF MILK PROTEINS BY IMMOBILIZED CELLS.

J.C. Vuillemard and J. Amiot
Departement De Sciences et Technologie Des Aliments
et Centre De Recherche En Nutriton
Universite Laval, Quebec, Que., Canada, G1K 7P4

ABTRACT

Protease production by Ca-alginate immobilized Serratia marcescens cells in a batch system was investigated. Various conditions of enzyme production by the immobilized cells in tryptone yeast medium were tested. Higher protease activity rates were obtained with 2.0% (W/V) starting cell content within the beads, and 10% (V/V) of beads inoculum. Aeration and addition of milk proteins in the medium induced the protease production. Cell immobilization was effective for elongation of the protease production half-life. As a result, the half-life of the protease production by immobilized cells was shown to be about 12 days against 5 days for submerged culture. Enzymatic hydrolysis of skim milk proteins was investigated in a batch reactor. Free amino groups liberated were linear with time for the first 40 hours.

KEYWORDS

Immobilized cells - Aliginate - Protease - Proteolysis - Milk proteins - Serratia marcescens.

INTRODUCTION

The use of proteolytic enzymes of the modification of food proteins such as solubilization of fish protein concentrate (Hevia et al., 1976), tenderization of meat, chill proofing of beer... is common. Enzymatic hydrolysis improves functional properties (Adler-Nissen et al.,1983) and nutritional value (Lacroix et al., 1983) of vegetable proteins.

Cheese-making is the larger application of proteolytic enzymes. Fortification of soft drinks and fruit juices (Ma et al., 1983), decrease of antigenicity properties of B-lactoglobulin, predigested dietary products (Hernandez and Asenjo, 1982) are other applications of enzymatic hydrolysis of milk proteins.

The final aim of our project is to produce from milk proteins, a predigested dietary product which can be ingested by hospitalized patients for paranteral hyperfeeding.

213

Usually, expensive preparations of purified enzymes are used. The use of an immobilized bacteria which produce exocellular proteases is a way to reduce the enzyme cost.

In previous work (Vuillemard and Amiot, 1985) we characterized a bacteria: Serratia marcescens which produces an exocellular protease. S.marcescens was immobilized for production of 2-Ketogluconic acid (Venkatasubramanian et al., 1978), L-isoleucine (Wada et al., 1979, 1980), L-arginine (Fujimura et al.,1984).

The purpose of this investigation is to study the effect of immobilization conditions and inducers on protease production and the enzymatic hydrolysis of skim milk proteins in a batch system by immobilized S. marcescens.

MATERIALS AND METHOD

Materials

Tryptone yeast and plate count agar were purchased form Difco Laboratories, Detroit, MI, USA. Sodium alginate was from Kelco Company, San Diego, Calif, USA. Calcium chloride and tri-sodium citrate were obtained from Fisher Scientific, Fair Lawn, NJ, USA. Sodium tetraborate was from Anachemia Chemicals, Montreal, Que., Canada. Hide Power Azure, Trinitrobenzenesulfonic acid, tris hydroxyaminomethane/HCl and leucine were from Sigma, Saint-Louis, MO, USA. Spray-dried skim milk (protein content = 35.90%) was a gift from Agrinove, Sainte-Claire, Que., Canada. Whey proteins concentrate (protein content: 34.99%) was from Sodispro Technologie, Saint-Hyacinthe Que., Canada. Antifoam was ATMOS 300 from Atkemix, Brantford, Ont., Canada. All other chemicals were analytical grade. Deionized water was used in all procedures.

For hydrolysis of milk proteins, the use of a stirred tank reactor will probably yield better results than a column fermentor, where plugging of milk proteins can occur, and better than a fluidized bed where gas bubbles in milk will produce too much foam. Therefore, we choose to work with a cytocultor.

Microorganisms and Immobilization Conditions

Culture. Serratia marcescens N-12 maintained at 4°C. on tryptic soy agar was used in these studies. Batch fermentations of immobilized S. marcescens were performed with Tryptone Yeast (TYE) medium (pH 7.0) in Erlenmeyer flasks at 30°C. under aerobic conditions [rotary shaker (150rpm)] or in a well mixed 2-1 cytocultor (Setric Genie Industriel, Toulouse, France) at 50 rpm.

Immobilization. S. mercescens was precultured 24 h in a flask under conditions described above. After centrifugation (15 min at 15 000 g), about 2 g wet cells (100 mL precultured broth) were suspended in 5 mL saline and mixed to 100 mL of a 2.0% (W/V) sodium alginate solution. The mixtue obtained was extruded through a syringe to a gently stirred 3% (W/V) CaCl$_2$ solution and hardened in this solution for 30 min. The resulting alginate gel beads (mean diameter of 4 mm) were thoroughly washed with saline. During reaction in TYE medium, 0.01M sodium chloride was added.

Analytic Methods

Determination of Cell Growth. Growth was measured by the aborbance of culture samples in a spectrophotometer(Hewlett Packard 8451) at a wavelength of 600 nm.

The number of living cells was estimated by dissolving 0.5 mL of gel beads in 4.5 mL tri-sodium citrate 1% (W/V) with shaking. The cell suspension obtained was serially diluted, and the number of viable cells was counted by the drop plate method and was expressed as the number of living cells per mL of gel.

Protease Assay.

Cell free supernatant fluids were obtained by centrifugation of 4.0 mL aliquots from medium at 15 000 g for 15 min at 0-4 °C.. The supernatant was filtered through a 0.45 μm. Millipore membrane. A filtrate of 1 mL was assayed for proteolytic activity by the procedure of Flyg and Xanthopoulos (1983), modified as follows. A sample containing 20 mg of Hide Powder Azure and 1 mL of enzyme dilution was made up to 5 mL with 4 mL 0.05M Tris-HCL pH 7.0 buffer solution. The tubes were incubated at 37 °C. for 2 h and the substrate maintained in dispersion by constant inversion (60 rev/min). The reaction was stopped by filtration through Whatman No. 1 filter. The absorbance of the filtrate, which represents a measure of the release of the blue dye covalently labelled to the protein substrate (denatured collagen), was read at 595 nm against a blank to which no enzyme was added. An increase in absorbance of 0.10 during 2 h incubation at 37 °C. was defined as one unit of proteolytic activity.

Analysis of Free Amino Groups. Free amino groups were determined automatically by the TNBS method of Palmer and Peters (1965) with a Technicon auto-analyzer (Technicon Corporation, Tarrytown,Nyn USA).

Absorbances of duplicate samples were converted to micromoles of free amino groups per milliliter of milk by a standard curve of leucine. Proteolysis was defined as the increase in the concentration of free amino groups per milliliter of milk (μm./mL).

RESULTS

Effect of Cell Content within Beads on Protease Production

10 mL of beads containing 0.2, 1.0, 2.0, 10.0 and 20.0% (W/V) wet cells were incubated at 30 °C. in 500 mL flask with 190 mL TYE medium (pH 7.0) on a rotary shaker (150 rpm). Figure 1 shows the protease activity produced. Except the 0.2% curve, the other cell contents did not show enzyme production difference in time. Maximum protease activity was reached by 2.0% (W/V) immobilized cells. Therefore, a cell content of 2.0% (W/V) was employed for further experiments.

216

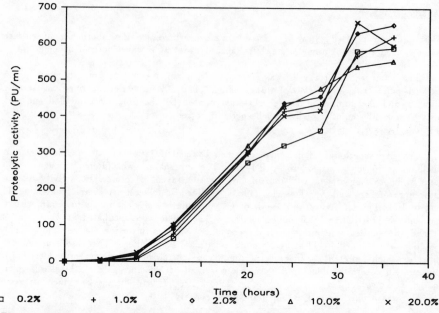

Fig. 1. Effect of cell content in calcium alginate beads on the protease production. Immobilization and reaction (5% inoculum) were carried out under standard conditions except cell content in the gel.

Table 1 shows cell concentrations within beads of 0.2, 1.0 and 2.0% (cell content) to increase to 10^{10} cells/mL of beads. This number seems to be the maximum cell concentration in beads. 10^{10} cells/mL of beads is the starting cell concentration at 10.0 and 20.0% cell content which stayed the same even after 36 h incubation.

Table 1. Immobilized cells concentration (number of cells/ml of gel) of various cell content within the beads.

| Time (hours) | Starting cell content in percent (w/V) | | | | |
	0.2	1.0	2.0	10.0	20.0
0	2.5×10^9	5.0×10^9	5.4×10^9	1.15×10^{10}	3.3×10^{10}
36	8.0×10^9	4.2×10^{10}	1.2×10^{10}	2.50×10^{10}	3.0×10^{10}

Effect of Beads Inoculum on Enzyme Production

Various volumes of beads were incubated in TYE medium to give a final volume of 200 mL with 1.0%, 5.0%, 10.0%, 20.0%, 30.0% and 50.0% (V/V) inoculum in 500 mL erlenmeyer flasks.

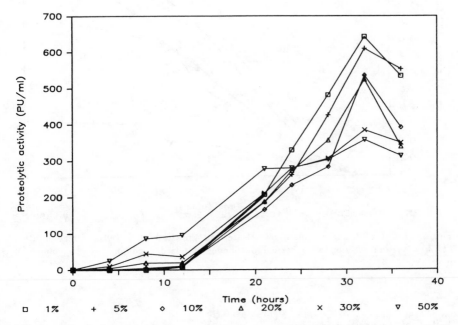

Fig. 2. Effect of beads inoculum on the protease production. Immobilization (2.0% w/v) and reaction were carried out under standard conditions except volume of inoculum.

In the first 24 h, protease activity was superior in flasks with larger inoculums(30 and 50%) as shown in Fig. 2. However, after 24 h, enzyme production increased as inoculum volume decreased. Greater enzyme activity with larger bead inoculum is explained by a greater release of enzymes from the beads. The volume of the medium and the nutrients decreases as bead inoculum increases. After 24 h incubation, nutrient concentrations in larger inoculum was too low for optimal enzyme rates. Low nutrient concentrations also limit growth of released cells. In the case of media with low inoculum, free cells grew and produced enzyme. This is shown by optical densities of the medium (Fig. 3). 1% and 5% of bead inoculum produced highest protease activity, but also highest free-cell concentration. On the other hand, 30% and 50% inoculum produced less free cells but less enzymes. A 10% bead inoculum was selected for further investigations.

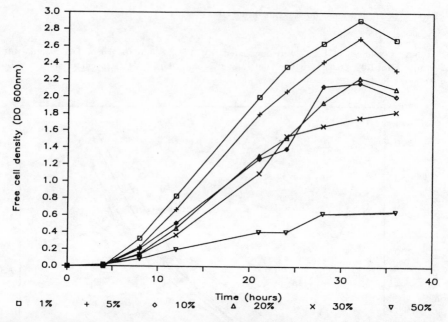

Fig. 3. Effect of beads inoculum on release and growth of free cells. Immobilization (2.0% w/v) and reaction were carried out under standard conditions except volume of inoculum.

Half-life of the Immobilized Cells

10% (v/v) immobilized cells were incubated in TYE medium (pH 7.0) at 30 ˚ C., 150 rpm. Samples were collected every 24 h for protease activity determination, then the gel was thoroughly washed with saline. The immobilized cells were then resuspended in the fresh reaction medium. This procedure was repeated many times (15 times). Ten percent (v/v) of a free cell culture was transfered every 24 h in fresh medium.

Figure 4 shows the protease production by immobilized and free cells after 24 h reaction for 15 days. The protease produced by immobilized cells increased with increasing use cycles and reached a maximum after six cycles. The protease activity at this time was about 640 proteolytic units/ml and 3 times higher than that of the submerged culture.

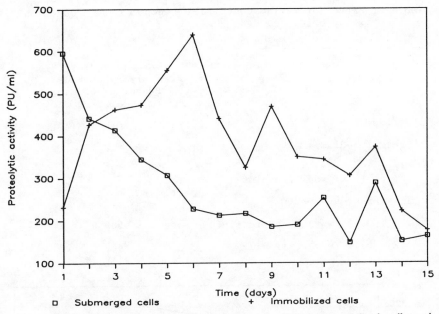

Fig. 4. Protease production by washed and calcium alginate immobilized cells under standard conditions.

The half-life of the immobilized cells, evaluated by extrapolating the line in Figure 4, and was about 12 days. Submerged cells showed higher productivities in the early cycles, but the activity rapidly decreased with increasing reaction cycles and the half-life was about 5 days.

Effect of Medium Constituents on Protease Production by Immobilized Cells.

As shown in Table 2, 24 h of reaction on saline or in a minimal glycerol-ammonia containing salt medium resulted in a very low protease production compared to the TYE nutritive medium. We conclude that protease is produced during the cell growth. The table shows the inducing effect of milk proteins on the enzyme productivity. In minimal medium, addition of 0.5% whey protein concentrate, multiplies the enzyme formation by 40.

Stimulative Effect of Proteins on Protease Production by Immobilized Cells.

The stimulative effect of 0.25%, 0.50%, 0.75% and 1.00% (w/v) whey protein concentrate and skim milk powder on protease production in TYE medium is shown in Figure 5. Enzyme production is increased by addition of skim milk or whey proteins. The enzyme rate is multiplied by 1.5 with addition of 1.00% skim milk and by 1.8 with 0.50% whey proteins.

Table 2. Effect of medium constituents on protease production by immobilized *S. marcescens.*

Medium	Proteolytic activity after 24 hours (PU per ml of medium)
Saline	0.1
Minimal medium	3.1
Minimal medium + 0.5% (w/v) whey proteins	126.7
Tryptone yeast medium	284.3
TYE + 0.5% (w/v) whey proteins	509.9
TYE + 0.5% (w/v) skim milk	409.5

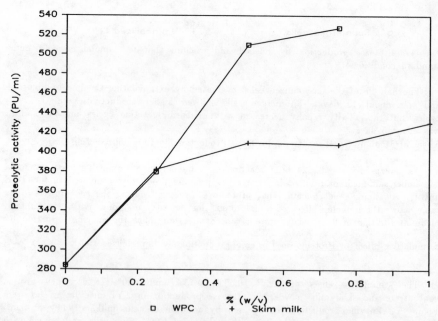

Fig. 5. Effect of skim milk powder and whey proteins concentrate on protease production under standard conditions.

Enzyme Production by Immobilized Cells in the Cytocultor.

   The immobilized cells (100ml of beads) were added to 900 ml TYE medium. The reaction was performed at 30°C. with controlled pH (7.0) and agitation (50 rpm) with or without aeration (air, flow rate: 10 1/h). With aeration, antifoam (0.5% v/v) was added in the TYE medium.

   Table 3 represents the protease activity after a 24 h reaction in the cytocultor under various conditions. Without aeration, agitation (50 rpm) is too low to permit sufficient oxygen transfer to the cells and protease activity in the medium is only 50 PU/ml.

Table 3. Protease activity in TYE medium after 24 hours of production into the cell culture fermentor under various conditions.

	Without aeration	With aeration (air: 10L/h)	With aeration and whey proteins concentrate (0.5%)
Proteolytic activity (PU/ml)	50.4	119.5	406.0

   The enzyme production is multiplied by 2 with aeration and by 8 with aeration and 0.5% (w/v) whey proteins.

Enzymatic Hydrolysis of Milk Proteins by Immobilized S. marcescens.

   The hydrolysis reaction was carried in the cytocultor. First, enzyme was produced by the beads in TYE medium with aeration under standard conditions for 24 h and reached a protease activity of 370 PU/ml. Then, reconstituted skim milk was added to give a final volume of 1.5 L containing 3.2% (w/v) proteins. Hydrolysis reaction was realised at 30°C., pH 7.0 with agitation (50 rpm) and without aeration during 48 h.

   Figure 6 shows the progress curve for the hydrolysis of skim milk proteins which represents the difference between free amino groups/ml of sample, less the free amino groups contained in milk before hydrolysis. The release of free amino groups from proteins was linear with time for the first 40 hours of reaction.
Proteolytic activity was diluted by milk addition to 150 PU/ml, but increased in the first 12 h of the reaction to reach 220 PU/ml. The activity remained stable for 40 hours (data not shown). Then, it decreased, and free amino group formation reached a stationary phase.

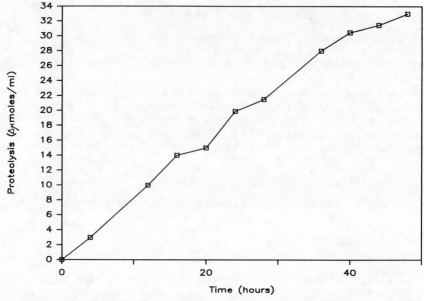

Fig. 6. Proteins hydrolysis (as the increase of the concentration of free amino groups per milliliter of milk) of reconstituted skim milk (3.2% proteins).

DISCUSSION

The protease production by immobilized <u>marcescens</u> seems to be independent of starting cell concentration in the range of 0.2% to 20.0% (w/v) (Fig. 1). Maximum enzyme rates increased as volumes of bead inoculum decreased (Fig. 2). This suggests that productivity depends on the ratio of bead volume/nutrients. With low bead inoculums, sufficient nutrients diffuse through the gel to the cells. On the other hand, with high inoculums, insufficiency in nutrients inside of the gels, results in less enzyme production by the immobilized cells. Furthermore, free cells in the medium increased as bead volumes decreased (Fig. 3). Leakage of cells from the gels may be more important in higher bead inoculums, however, free cell densities which may produce proteases are superior for lower inoculum concentrations. This may be due to the larger amount of nutrients available.

These observations did not show the obviousness of using immobilized cells for protease production. However, the main advantage of immobilized cells compared to free cell is their stability in time (Fig. 4). The half-life calculated was 12 days for immobilized cells against 5 days for submerged cells. Improvement of protease production half-life by immobilized mycelia was reported by Kokubu et al., 1981. These authors further improved the half-life by sterilization of the gel surface.

223

Proteins induce protease production by free cells of S. marcescens. In previous studies (Vuillemard and Amiot,1986), we showed the inducing effect of gelatin and casein on protease production. Murakami et al., 1969, observed a stimulating effect of α-lactalbumin on protease formation. Therefore, we tested the effect of milk proteins and we observed an induction of enzyme productivity by immobilized cells (Fig. 5), not only in nutritive medium but also in minimal medium (Table 2). This last medium, with milk proteins as inducer may separate protease production and cell growth.

This preliminary study on protease production by immobilized S. marcescens allows us to define immobilization and operating conditions of a 10% (v/v) bead inoculum with a starting concentration of 2.0% (w/v) wet cells for further investigations. Milk proteins (Table 2, Fig. 5) and aeration (Table 3) have a stimulating effect on protease formation.

Gentle agitation and the rounded wall of the reactor did not destroy the gel. We used the same gel for 10 days without observed bead destruction. Milk proteins are too large to diffuse into the gel and to be hydrolysed inside by immobilized cells. Therefore, we used a microorganism which excretes an exocellular protease in the medium. The first steps of hydrolysis occurred outside the beads, then hydrolysates of low molecular weight could diffuse into the gel for further hydrolysis. The calcium alginate gels provides little barrier to diffusion of substrates up to a molecular weight of 5,000 (Kierstan and Bucke, 1977). In further studies, molecular weight distribution of the hydrolysates will be determined.

REFERENCES

1. Adler-Nissen, J., Erickson, S. and Olsen, H. S.; Improvement of Functionality of Vegetable Proteins By Controlled Enzymatic Hydrolysis; Plant Proteins for Human Foods 207-219, 1983.

2. Flyg, C. and Xanthopoulus, K. G.; Insect Pathogenic Properties of Serratia marcescens. Passive and Active Resistance to Insect Immunity Studied with Protease - Deficient and Phage Resistant Mutants. J. Gen. Microb. 123: 453-464, 1983.

3. Fujimura, M., Kato, J. and Chibata, I.; Continuous Production of L. Arginine Using Immobilized Growing Serratia marcescens Cells.; Appl. Microbiol. Biotechnol. 19: 79-84, 1984.

4. Hevia, P., Whitaker, J. R. and Olcott, H. S.; Solubilization of a Fish Protein Concentrate with Proteolytic Enzymes. J. Agric. Food Chem.; 24: 2, 383, 1976.

5. Hernandez, R. and Asenjo, J. A.; Production of an Enzyme Hydrolysate of Skim Milk Lactose and Proteins.; J. Food Sci.; 47: 1895-1898, 1982.

6. Kierstan, M. and Budke, C.; The Immobilization of Microbial Cells, Subcellular Organelles, and Enzymes in Calcium Alginate Gells.; Biotechnol. Bioeng.; 19: 387-397, 1977.

7. Kokubu, T., Karube, I. and Suzuki, S.; Protease Production by Immobilized Mycelia of Streptomyces fradiae Biotechnol. Bioeng.; 23: 29-39, 1981.

8. Lacroix, M., Amiot, J. and Brisson, G. J.; Hydrolysis and Ultrafiltration Treatment to Improve the Nutritive Value of Rapeseed Proteins. J. Food Sci.; 48: 1644-1645, 1983.

9. Ma, C., Anantea, C. and Nakai, S. Production of Non Bitter, Desalted Milk Hydrolystate for Fortification of Soft Drinks and Fruit Juices. J. Food Sci.; 48: 897-899, 1983.

10. Murakami, M., Fukunaga, K., Matsuhashi, M. and Ouo, M. Stimulative Effect of Proteins on Protease Formation by Serratia sp. Bioch. Biophys. Acta.; 192: 378-380, 1969.

11. Palmer, D. W. and Peters, T.; Simple Automatic Determination of Amino Groups in Serum/Plasma Using Trinitrobenzene Sulfonate.; Automation in Analytical Chemistry, Technicon Symposia; 324-327, 1965.

12. Venkatasubramanian, K., Constantinides, A. and Vieth, W. R.; Enzyme Eng. 3: 29, 1978.

13. Vuillemard, J. C. and Amiot, J. Hydrolysis of Food Proteins by Bacteria. Microb. Alim. Nutr. 3: 333-343, 1985.

14. Vuillemard, J. C. and Amiot, J. Induction, Repression and Characterization of a Protease by Serratia marcescens. Submitted for publication, 1986.

15. Wada, M., Uchida, T., Kato, J. and Chibata, I. Continuous Production of L. Isoleucine Using Immobilized Growing Serratia marcescens Cells. Biotechnol. Bioeng. 22: 1175-1188, 1980.

RECENT PROGRESS IN THE IMMOBILIZATION OF β-GALACTOSIDASE

Yong K. Park and Glaucia M. Pastore
Universidade Estadual de Campinas,
Faculdade de Engenharia de Alimentos (UNICAMP)
Campinas, 13100, SP. Brasil

ABSTRACT

This article is a review about different methods of immobilizing β-galactosidases and the various characteristics which have appeared in recent competitive journals. Finally, these immobilized enzymes were compared with a preparation of immobilized Scopulariopsis β-galactosidase developed in this laboratory.

KEYWORDS

Lactose, Milk, Whey, β-Galactosidase, Immobilization, Hydrolysis.

INTRODUCTION

Considerable attention has been focused on the use of β-galactosidase, lactase (β-galactoside galactohydrolase, EC. 3.2.1.23) for the hydrolysis of lactose in dairy products, from the view-points of the full utilization of whey and the removal of lactose from the milk due to its low solubility and physiological problem for the lactose intolerant. In order to use the enzyme efficiently for the practical hydrolysis of latose in milk and whey, immobilization of the enzyme is superior to the soluble enzyme. A number of papers on the immobilization of lactase by various methods have appeared in the last two decades[9]. The objective of this review is to summarize recent progress in the immobilization of lactase and its application.

COMPARATIVE STUDIES OF IMMOBILIZED B-GALACTOSIDASE

Immobilization of Aspergillus Oryzae Lactase on Regenerable Affinity Chromatography Support (RACS) Attached to Sepharose. [3]

The synthesis of a regenerable hydrocarbon arm on crosslinked Sepharose was achieved according to the method described in reference 3. One volume of RACS was suspended in two volumes of dioxane, and dicyclohexylcarbodiimide and N-hydroxysuccinamide added at concentrations of 0.1 M. The reaction was allowed to proceed for 90 minutes at room temperature and then the activated resin was washed thoroughly with dioxane, absolute methanol, and dioxane. The slightly moist gel was suspended in a 0.067 M phosphate buffer, pH 6.5 which contained 40 mg lactase/mL packed gel, and the coupling reaction was allowed to proceed overnight at 4 °C. The resin was then filtered and resuspended in fresh buffer containing 0.2 M glycine. After two hours the resin was washed with ten volumes of buffer, resuspended, and stored at 4 °C until further study.

Both the soluble and immobilized fungal lactase preparations possessed the same pH optimum range, 4.5 - 5.0. Since the lactase possessed approximately 50% of its maximum activity at pH 6.5, it could possibly be used for the hydrolysis of sweet whey or milk at neutral pH, as well as at its acid pH optimum of 4.5.

At pH 4.5, the soluble enzyme possessed maximum activity at 60 °C, while the immobilized preparations possessed maximum activity at 55 °C. At pH 6.5 the soluble and immobilized enzyme preparations assayed under identical conditions in a batch process showed identical thermal profiles, with maximum activity at 55 °C, while the packed-column preparation possessed maximum activity at 60 °C.

When immobilized enzyme was placed in a packed-bed reactor, the effect of temperature on the activity was altered, as reflected by a marked decrease in the thermodynamic parameters of activation at both pH 4.5 and 6.5. Upon immobilization there was also a dramatic increase in the apparent thermal resistance of the lactase, and the mean half-life at 50 °C was increased from 7.2 to 13 days at pH 4.5 and from 3.8 to 16 days at pH 6.5.

Immobilization of Yeast Cells Containing Lactase. [1]

More recently Saccharomyces anamensis which contained β-galactosidase were immobilized by entrapping the cells into agar-agar. One milliliter of cell suspension (100 mg wet weight) in phosphate buffer was added to 5.0 mL of molten agar prepared by melting 125 mg of agar in 5.0 mL of distilled water, and then cooled to 45 - 50 °C and shaken thoroughly. This cell-agar mixture was then cast into bead shape by injecting it into an ice cold toluene-chloroform (3:1) mixture. The beads were then washed repeatedly with phosphate buffer.

Polyacrylamide gel lattice is widely used for enzyme immobilization because of its numerous advantages. Unfortunately, a low activity yield is obtained when enzymes are entrapped in a polyacrylamide gel matrix. β-galactosidase from Lactobacillus bulgaricus was immobilized on polyacrylamide gel, and the maximum activity yield was only 31%[6]. The factors responsible for the low activity yield of entrapped β-galactosidase could be the denaturation caused by acrylamide and the oxidation of essential sulfhydryl groups to -S-S bonds by potassium persulfate. It appears worthwhile to try the addition of protective agents such as substrate, inhibitors, reducing agents etc. during entrapment, to increase the activity yield of immobilized β-galactosidase.

Immobilization of β-Galactosidase from Bacillus Circulans on Duolite. [7]

   The enzyme is characterized by a considerably higher activity towards lactose in skim milk as good as that at pH 6.0 indicating that this enzyme is suitable for the treatment of lactose in milk. The respective temperature optimum for the hydrolysis of lactose in buffer solution and in skim milk were 60 and 65°, when incubated for 10 minutes. The time course for the hydrolysis of lactose in the assay buffer at 40°C using β-galactosidase from B. circulans was examined. As the reaction proceeded, not only glucose and galactose, but also considerable amounts of tri- and tetrasaccharides were produced. Trace amounts of oligosaccharides with higher degrees of polymerization were also detected. Far more trisaccharide was produced than glucose or galactose, particularly in the initial stages of the reaction. Time courses for the hydrolysis of lactose with galactosidases from E. coli and S. lactis were compared with the results for B. circulans. β-galactosidase from B. circulans produced considerably larger amounts of tri- and tetrasaccharides than the other enzymes, with a maximum yield at 40% conversion of lactose. However, very little oligosaccharides could be detected at conversions > 80%. On the other hand, S. lactis β-galactosidase produced a maximum amount of trisaccharide in a conversion range between 45 and 85%. E. coli galactosidase produced only a small amount of trisaccharide over the whole conversion range tested.

   Duolite ES-762 was used as a support for the immobilization in the presence of glutaraldehyde. The adsorption equilibria between the enzyme concentration in solution and the amount of enzyme adsorbed per gram of wet Duolite resin showed that the yield of immobilization decreases as the amount of enzyme adsorbed increases. There is very little difference in activity between the enzyme adsorbed on the resin with or without glutaraldehyde treatment. An intraparticle diffusional effect might be largely responsible for the decrease in the yield of immobilization. The activities of the immobilized enzyme derivations on various supports, which were prepared by adsorption at an equilibrium concentration of 130 units mL^{-1} are compared. The enzyme immobilized on Duolite ES-762 showed a considerably higher activity than that on other supports such as Dowex MWA-1 and sintered alumina.

   The pH-activity profile for galactosidase immobilized on Duolite ES-762 had no distinct peak. The maximum activity was observed in the pH 6.5 to 8.2 region, showing that immobilization caused a shift in pH optimum towards the alkaline region. The time course for the hydrolysis of lactose in the assay buffer using β-galactosidase immobilized on Duolite ES-762 was examined, and it was found that in the initial stage of the reaction, tri- and tetrasaccharides were produced to a much greater extent than with the free enzyme, possibly due to the effect of intraparticle diffusion.

   During the continuous hydrolysis of lactose, the immobilized enzyme was reversibly inactivated, probably due to the accumulation of oligosaccharides in the gel. The inactivation was reduced when the continuous reaction was operated at a high percent conversion of lactose in a continuous stirred tank reactor (CSTR). When the reaction was carried out in a CSTR with a percent conversion of lactose > 70%, the half-life of the immobilized enzyme was estimated to be 50 and 15 days respectively at 50 and 55° C. It was found that oligosaccharides, which were produced as a by-product in the hydrolysis of lactose, prompted a reversible inactivation of the immobilized enzyme.

E. coli β-Galactosidase Immobilized on Photoactivable Chitosan. [5]

The chitosan was successively washed with 1 M ammonia and water and then dried. One equivalent of triethylamine (0.8 mL) and a 10-fold excess of 4-azido-3,5-dichloro-2,6-difluoropyridine (2.2 g) were added to a mechanically stirred suspension of chitosan (1 g) in 20 mL dimethylformamide. The reaction was then run at a temperature of 80°C for 72 hours. After stopping the reaction, the resulting resin was isolated by filtration and successively washed with methanol and methylene chloride, before drying under vacuum. Immobilization of E. coli β-galactosidase (970 U mg^{-1}) was performed as follows; 100 mg photoactivable chitosan was washed with 0.1 M sodium phosphate buffer, pH 7.3, and then mixed with β-galactosidase (1 mg) in 2 mL 0.01 M phosphate buffer, pH 7.3. The resulting mixture was irradiated (above 300nm) for different time periods. The buffer used had been deoxygenated with a stream of nitrogen. The enzyme was covalently bound into the matrix. The coupling of the protein to the matrix is effected after photochemical generation of highly reactive nitrene, which also reacts competitively with the solvent. Consequently, several structural parameters of the matrix must influence the binding of the protein and the activity of the bound enzyme.

The pH optimum for β-galactosidase immobilized on photoactivatable chitosan was examined, and it was found that maximum activity of the immobilized enzyme is shifted towards a basic pH as compared to the soluble enzyme. The operational stability of the immobilized enzyme was tested in a continuous packed-bed reactor using lactose as substrate. At a flow rate allowing conversion of 50% of the substrate, no decrease in the enzyme activity was observed during 12 days.

Immobilization of E. coli Lactase in Hen's egg. [4]

E. coli lactase was immobilized in hen's egg white (5 mL) using glutaraldehyde (2 %) as the crosslinking agent. The egg white (5 mL) was mixed thoroughly with the enzyme (5 mg in 0.5 mL water) and then treated with glutaraldehyde to a final concentration of 2%. The mixture was stirred well and allowed to stand for 2.5 hr at 25°C. The hard gel obtained was then shattered by passing through a syringe needle and washed with water, free excess glutaraldehyde, and enzyme. The immobilized enzyme was washed free of excess lysine with water, and stored as an aqueous suspension at 4°C.

Both soluble and immobilized E. coli lactase exhibited optimal activity at pH 6.8, although there was a broadening of the activity curve towards the acidic side for the immobilized enzyme. Immobilization caused an increase in the optimal temperature, from 40 to 50°C. The enzyme was also rendered more thermostable. The immobilized preparation showed a higher km for the substrates. The extent of the enzyme inhibition by galactose was reduced upon immobilization. The stabilities towards inactivation by heat, urea, gamma irradiation, and protease treatment were enhanced. The bound enzyme, as tested in a batch reactor, could be used repeatedly for the hydrolysis of milk lactose. The possible application of this system for small-scale domestic use has been suggested.

Immobilization of Thermus Strain 4-1 A Lactase on Pore Glass. [2]

An inducible β-galactosidase from an extremely thermophilic bacterium was immobilized by using 1.5 g of controlled pore glass, which was silanated with λ-aminopropyltriethoxysilane and then treated with 10% glutaraldehyde. Fifty milliliters of a β-galactosidase solution (2.86 mg/mL in 0.04 Tris-Hcl, pH 9) were added, and the suspension incubated at 40°C with stirring for 72 hours. The immobilized preparation was washed exhaustively.

The enzyme was very stable at pH 8 in dilute buffer solutions at temperatures of up to 90°C. Sodium dodecylsulphate (0.1 %) and 10 mM mercaptoethanol slightly reduced the stability at 75°C, while 0.5 % NaCl increased thermostability at 80°C. Immobilization of the enzyme by glutaraldehyde cross-linking to controlled-pore glass resulted in apparent inhibition of 64 % of the bound enzyme. Such inhibition is normally attributed to attachment of the enzyme in an orientation in which the active site is sterically hindered. Little alteration of the pH-activity profile and no change in product inhibition was observed after immobilization.

Immobilization of Scopulariopsis β-galactosidase.

Recently, we have isolated a strain of Scopulariopsis sp. from soil, which produced high extracellular β-galactosidase on wheat bran culture medium. Maximum hydrolysis of this enzyme occured at pH 4 and 5, and for ONPG this was at pH 3.6 to 5. The optimum temperature for the hydrolysis of lactose was 50 - 60°C and maximum hydrolysis of ONPG occured. Paper chromatograms of lactose hydrolysates by the enzyme indicated that the enzyme hydrolyzed lactose almost completely to glucose and galactose.

A comparison of Scopulariopsis lactase with existing microbial lactases for dairy processing is shown in Table 1.

Microorganism	pH optimum	Optimum temperature	Location of enzyme
Aspergillus oryzae	4-5	55-60	Extracellular
Aspergillus foetidus	3.5-4.0	66-67	Extracellular
Aspergillus niger	3.5-4.5	55	Extracellular
Kluyveromyces fragilis	6.3-6.5	35-45	Cell bound
Scopulariopsis sp.	3.6-5.0	50-65	Extracellular
Escherichia coli	6.8	40	Intracellular
Bacillus circulans	6-6.5	60-65	Intracellular

Table 1. Comparison of Scopulariopsis lactase with other microbial lactase

Scopulariopsis β-galactosidase is highly thermostable and very active in an acid pH range as compared to yeast and bacterial lactases. The combination of high temperature and acid pH for optimum activity of lactase is desirable for acid whey processing, although a low pH is not suitable for treating milk. Therefore, it is apparent that Scopulariopsis lactase is applicable to the cottage cheese operation, and does offer a potential solution for the serious problem of acid whey disposal. These experimental conditions can also limit microbal growth during hydrolysis.

The enzyme was extracted by mixing water to culture of Scopulariopsis sp., on wheat bran and concentrated by percipitating with ethanol from extract. The concentrated enzyme was immobilized on both DEAE-cellulose and phenolformaldehyde resin (Duolite S-761) with or without glutaraldehyde . HI-FLO Selectacel DEAE-cellulose obtained from Brown Company, New Hampshire, USA, mesh size 30 - 60 was used. Fourteen grams of DEAE-cellulose was swelled, activated, and equilibrated with 0.05 M acetate buffer, pH 5.0. Two g of the enzyme preparation was dissolved in 100 mL of acetate buffer, pH 5, and centrifuged to remove insoluble materials. The activated DEAE-cellulose (82 g) was added to the enzyme solution. After 2 hr standing with occasional stirring of the mixture, glutaraldehyde (final concentration 2.5%), was added, and allowed to stand overnight at 4°C. The immobilized enzyme was washed thoroughly with acetate buffer. Alternatively, 20 g of a phenol-formaldehyde resin (137 g as weight), obtained from Diamond Shamrock Corporation, USA, registered as Duolite S-761, 40 - 80 mesh was also used for immobilization of the enzyme as described above.

Table 2 shows comparison for immobilization of Scoplariopsis lactase on both DEAE-cellulose

Treatment	Uncoupled enzyme activity in supernatant (units/ ml)	Immobilized enzyme activity (units/g resin)
Adsorption on DEAE-cellulose	10.6	0.90
Immobilization of the enzyme on DEAE-cellulose with glutaraldehyde	8.3	1.20
Adsorption on Duolite	7.7	0.15
Immobilization of the enzyme on Duolite with glutaraldehyde	5.4	0.20

Table 2. Comparison of activities of β-galactosidase immobilized on DEAE-cellulose or Duolite by adsorption and by both adsorption and covalently bonding with glutaraldehyde.

and Duolite S-761. The enzyme activity remaining in the supernatant (uncoupled enzyme) after immobilization was 8.3 and 5.4 units for DEAE-cellulose and Duolite respectively. These results indicate that 58.5% of the enzyme was adsorbed on DEAE-cellulose while Duolite adsorbed 73% of the enzyme. The immobilized enzyme on DEAE-cellulose showed 6 times more activity than that on Duolite, although Duolite adsorbed more enzyme proteins. It can be seen that the combination of adsorption and covalent bond resulted in the slightly higher activity than adsorption only.

The effect of NaCl on the enzyme immobilized on DEAE-cellulose and Duolite was investigated, and the loss of the enzyme activity from immobilized on DEAE-cellulose without glutaraldehyde treatment was observed at over 0,1 M NaCl and was 92% at 0.3 M. On the other hand, the activity loss of immobilized enzyme with glutaraldehyde treatment was quite slight at 0.3 M of NaCl and was 40% even at 0.7 M. It was also found that the enzyme immobilized on Duolite both with or without glutaraldehyde treatment was not released at 1.5 M of NaCl solution. It is obvious that the enzyme was strongly bound to Duolite, although the yield of the immobilized enzyme was far less than that on DEAE-cellulose.

The pH optimum of both types of the immobilized enzyme was found to be between 4 and 5. Thermal stability of the immobilized enzyme was investigated by incubating the test tubes containing a suspension of 1 g wet immobilized enzymes in 1 mL of acetate buffer at 50, 55 and 60 ° C, respectively. After 80 hr incubation, the remaining activities of the immobilized enzyme were 14% and 80% at 60 and 55 ° C, respectively. At 50 ° C, there was no loss of enzyme activity. In continuous operation of lactose hydrolysis in whey. Both immobilized enzymes were packed in a jacketed column (2.0 x 1.8 cm) respectively and acidic whey was continuously applied to the columns to pass down at a rate of 50 mL/hr. The temperature of the columns was controlled by circulating water at 55 ° C. As shown in Figure 1, the column with β-galactosidase immobilized on DEAE-cellulose in presence of glutaraldehyde maintained the enzyme activity for 30 days with 25% loss of the activity, while the column with the enzyme immobilized on Duolite maintained hydrolysis of lactose without loss of the enzyme.

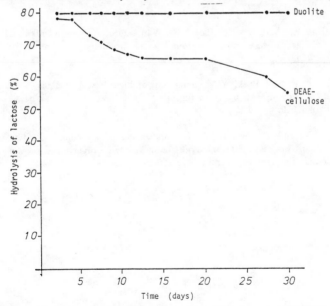

Fig. 1. Continuous operation for β-galactosidase immobilized on both DEAE-cellulose in presence of glutaraldehyde and Duolite in column.

REFERENCES.

1.   M. Banerjee, A. Chakrabarty, and S.K. Majumdar. Immobilization of yeast cells containing β-galactosidase.; Biotechnol.Bioeng.; <u>24:</u> 1839-1850, 1982.

2.   Cowan, D.A., Daniel, R.M., Martin, A.M. and Morgan, H.W.;Some properties of a B-galactosidase from an extremely thermophilic bacterium. Biotechnol.Bioeng.; <u>26:</u> 1141-1145, 1984.

3.   Friend, B.A. and Shahani, K.M.; Characterization and evaluation of aspergillus oryzae lactase coupled to a regenerable support.; Biotechnol.Bioeng.; <u>24:</u> 329-345, 1982.

4.   Kaul, R., D'Souza, S.F. and Nadkarni, G.B.; Hydrolysis of milk lactose by immobilized β-galactosidase-hen egg white powder.;Biotechnol.Bioeng.; <u>26:</u> 901-904, 1984.

5.   Leonil, J., Sicsic, S., Braun, J. and Goffic, F.;A new photoactivable support for protein immobilization: fixation of β-d.galactosidase on modified chitosan.; Enzyme.Microb.Technol.; <u>6:</u> 517-521, 1984.

6.   Makkar, H.P.S., Sharma, O.P. and Dawra, R.K.; Effect of reagent for polyacrylamide gel formation on β-D.galactosidase.;Biotechnol.Bioeng.; <u>25:</u> 867-868, 1983.

7.   Nakanishi, K., Matsuno, R., Torii, K., Yamamoto, K. and Kamikubo,T.; Properties of immobilized B-d.galactosidase from <u>bacillus</u> <u>circulans.</u>; Enzyme.Microb.Technol.; <u>5:</u> 115-120, 1983.

8.   Pastore, G.M. and Park, Y.K.; Screening of high β -galactosidase - producing fungi and characterizing the hydrolysis properties of a selected strain.;J. Food Sci.; <u>44:</u> 1577-1579, 1979.

9.   Park, Y.K. and Pastore, G.M.; Immobilization of scopulariopsis β-galactosidase to deae-cellulose with glutaraldehyde and some properties of the preparation obtained.;J. Ferment. Technil.; <u>59:</u> 165-168, 1981.

IMMOBILIZATION OF LACTASE AND INVERTASE ON CROSSLINKED CHITIN

A. Illanes, M.E. Zúñiga, R. Chamy and M.P. Marchese
Universidad Católica de Valparaíso. Casilla 4059, Valparaíso, Chile.

ABSTRACT

Chitin has been selected as a suitable support for lactase and invertase immobilization with glutaraldehyde. Different protocols of immobilization were evaluated for both enzymes in terms of immobilization yield, charge capacity and operational stability. At optimum conditions, immobilization efficiency was 46% and 71% and catalyst operational half-life was 10 and 26 days for lactase and invertase respectively.

Immobilized invertase followed simple Michaelis-Menten Kinetics; lactase was competitively inhibited by galactose and slightly activated by glucose. Both catalysts were essentially free of external diffusional restrictions.

Immobilized enzymes have been tested in long-term continous operation in fixed-bed reactors. For the case of invertase, results agree reasonably with a simple model based on piston flow. Modeling of reactor operation with immobilized lactase is underway.

Keywords: Immobilized lactase, immobilized invertase, crosslinked chitin, sucrose hydrolysis, lactose hydrolysis, whey hydrolysis.

INTRODUCTION

Invertase (B-D fructofuranoside fructohydrolase E.C.3.2.1.26) is an enzyme used traditionally in the production of invert sugar for the food industry[52]. Although at present its market is limited[42], an increase is expected as invert sugar can be used to produce high fructose syrups with the already developed technology of

233

glucose isomerization and fructose separation. Immobilization of invertase has been extensively studied over a wide variety of supports like polyacrylamide gel[22], porous glass[33], cellulose fibers[31], ion exchangers[15,53], krill chitin[47], PVA membranes[20], activated clay[25], bentonite[48] and also to methacrylate, styrene and cellulose, via its carbohydrate moiety[32]

Lactase (B-D galactoside - galactohydrolase E.C. 3.2.1.23) is being actively studied for the production of hydrolyzed milk[14,30,40], whey[7,14,44,49], and whey permeate[7,23,49]. Several pilot plants are already in operation[4,7,30,40,44] and some commercial ventures are underway [1,3,5,43]. The market for hydrolyzed milk relies in the production of diet milk and dairies for lactose intolerants, and the production of dairy concentrates[14]. Whey is very often wasted and, having a high BOD[24], its upgrade is also interesting from a waste treatment stand point. Hydrolyzed whole whey can be used as a fermentation medium for ethanol production and eventually other metabolites[14]. Also it may be used in the food industry as a sweet protein premix[7] or else as a caloric supplement in nutritional programs in developing countries[11]. The market for deproteinized (demineralized) whey relies mainly in the production of syrups for assorted applications in the food industry[7,14].

Although several sources of lactase are available, only those from yeast and fungi are worth considering for process development. Fungal lactases have acid pH optima and are therefore recommended for whey or permeate processing. Yeast lactases are neutral and are a choice for milk and sweet whey hydrolysis. Recently, immobilized whole cells, rich in lactase, have also been considered for whey processing[6,51]. It is apparent that immobilized enzyme technology is especially applicable to these processes and, therefore, much effort has been devoted to study lactase immobilization. Comprehensive reviews on the subject have been published recently[10,13]. Immobilization system and reactor configuration are very much dependant on substrate properties. Protein-containing substrates, such as milk or whole whey, are preferably treated in hollow-fiber type reactors with the enzyme either free in solution or immobilized in the outer membrane sponge layer[16,17,29,38,39,40]. Protein-free substrates, like whey permeate are better processed in packed-bed reactors with lactase immobilized to solid supports[5,7,23,28 36,44,54]

Among several others, we have selected crosslinked chitin as a suitable support for the immobilization of microbial enzymes[19,34]. Preliminary work with a partially purified yeast invertase[19], prompted us to continue this work with a crude invertase preparation and to test the behavior of the catalyst in continous packed-bed reactors. The methodology has been extended to a more complex

system in which a commercial fungal lactase was immobilized to crosslinked chitin.
The immobilization procedure was optimized in terms of enzyme to chitin (E/S) and
glutaraldehyde to chitin (G/S) ratios, for protocols in which different sequences
of addition of reagents and incubation times were tested. Optimum conditions were
defined as those who maximized the charge capacity of the catalyst at an acceptable
enzyme immobilization yield. Temperature and pH profiles were determined for the
selected catalyst and kinetic parameters were determined at optimum conditions.
Kinetic models, based on piston flow and appropriate kinetic expressions were
developed for immobilized   invertase and lactase. Long-term operations were
carried out using lactase and demineralized whey permeate for the case of lactase,
and commercial beet sugar for the case of invertase. Periodic cleaning of the
catalysts was essential to maintain a low rate of enzyme deactivation.

MATERIALS AND METHODS

Chitin flakes were obtained from waste shrimp shells by a procedure similar
to that reported by Muzzarelli[35], or else purchased to SIGMA Chemical Co. (St.
Louis, Mo, USA). A crude yeast invertase preparation (31 IU/mg protein) was
produced as reported previously[18]. Among commercial lactases available, ENZECO
fungal lactase from Aspergillus oryzae (Enzyme Development Corporation, New York,
USA) was selected for giving the lowest price per IU under our conditions of
operation. The enzyme had a specific activity of 71.1 IU/mg protein (18.5 IU/mg
solids). Glutaraldehyde in 25% solution was purchased to SIGMA Chemical Co.
Commercial, refined beet sugar, less than 0.01% glucose (IANSA, Chile) and USP
lactose were used as substrates. Demineralized whey permeate (35 to 40 g/l of
lactose) was kindly supplied by Dos Alamos S.A. (Santiago, Chile). Other
reagents were analytical grade and purchased to Merck (Darmstadt, West Germany).

Invertase and lactase activities were expressed in international units (IU)
at 53 °C, pH 4.5 and 55 °C, pH 4.0 respectively. Hydrolyzed sucrose was calculated
from the amount of reducing sugars produced, determined according to Folin-Wu[9].
Hydrolyzed lactose was determined from the amount of glucose produced, using the
glucostat procedure (glucose kit N° 510, SIGMA Chemical Co.). Protein was
determined according to Lowry et.al[27].

Immobilization was carried out using four different procedures, as indicated
in figure 1. Three of them involved glutaraldehyde as a cross-linking agent. After
each stage, the solid catalysts were washed with distilled water, 0.1M acetate
buffer pH 4.5 or 4.0 and 2M NaCl to eliminate any unbound protein. In the latter

stage at 4°C, no variations were observed in all cases after 10 hours of contacting, a standard time of 12 h being adopted.

Packed-bed reactors(d_r= 1,9 cm; h = 17 cm) were jacketed pyrex-glass columns, with bed height to diameter ratios from 7 to 8. They were always operated upflow with no recirculation. Reactor and connections were sanitized prior to use and substrates were sterilized at 121°C for 20 minutes.

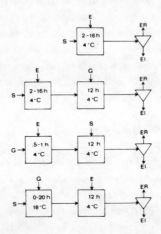

Fig.1 Immobilization procedures for invertase and lactase

RESULTS AND DISCUSSION

Best results, both for invertase and lactase, were obtained with immobilization procedure 4, in which chitin was pretreated with glutaraldehyde for 20 h at 18°C, then contacted with the enzyme for about one hour at 18°C, stored for 12 h al 4°C. and finally washed as indicated. Y values for both enzymes are presented in Table 1 for some of the procedures shown in figure 1. In each case, results correspond to the higher Y values obtained at varying E/S and G/S.

Table 1.   Immobilization yield for different immobilization procedures

Immobilization Procedure   . Y (%)

	Invertase	Lactase
1	26	21
2	--	10
3	46	18
4 t_1 = 20 h	70	46 (44)*
t_1 = 0 h	-	19 (59)

* Values in parenthesis represent results obtained by Stanley et.al[45].

Figure 2 present the Y and C values as a function of G/S for invertase and lactase at their corresponding E/S optima, as indicated in Table 2.

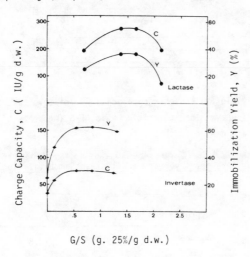

G/S (g. 25%/g d.w.)

Fig.2   Immobilization yield (Y) and charge capacity (C) of invertase and lactase as a function of glutaraldehyde to chitin ratio, G/S, (g 25% / g d.w.), at optimum enzyme load to chitin ratio, E/S.

238

Figure 3 presents the Y and C values as a function of E/S for invertase and
lactase at their corresponding G/S optima. As can be seen, maximun C was obtained
in both cases at suboptimal Y. However, since C will determine reactor size for
a given set of operating conditions, we chose the parameters that maximized C
at an acceptable Y. Selected values are summarized in Table 2.

C values were much higher for lactase than for invertase, while the opposite
held true for Y values.

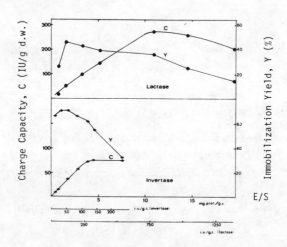

Fig.3. Immobilization yield (Y) and charge capacity (C) of invertase and lactase
as a function of enzyme load to chitin ratio, E/S (mg.protein / g d.w.),
at optimum glutaraldehyde to chitin ratio, G/S.

Table 2 Selected parameters for invertase and lactase immobilization

Enzyme	G/S g 25%/g d.w.	E/S mg prot/g d.w.	E/S IU/g d.w.	C IU/g d.w.	Y (%)
INVERTASE	0.55	3.82	119	73.8	62
LACTASE	1.36 (0.34)*	10.85	770 (404)	273.4 (143)	36 (36)

* Values in parenthesis represent results obtained by Stanley et.al[45].

239

In the case of invertase, C increased up to a contacting protein load of about 5 mg / g d.w., no decrease being observed at higher loads. This means that the support simply became saturated around that figure, as suggested by the increasing values of ER obtained thereon . For lactase, however, C increased up to a higher protein load of about 10 mg/g d.w, which can be attributed to the more heterogeneous nature of the invertase yeast extract as compared to the extracellular lactase. However in this latter case, a significant decrease in C was observed at higher protein loads with no concomitant increase in ER. This means that the support was binding more protein but, maybe due to oversaturation, enzyme molecules became clogged and activity of the catalyst was significantly impaired at these higher loads. Data for lactase are compared in figure 4 with those calculated from the results reported by Stanley et. al[45] with a similar system. As opposite to our case, they obtained higher Y values when all reagents were added at the same time (t_1=0 in procedure 4, see Table 1). Under similar conditions as in procedure 4, however, our results are comparable in terms of Y values, but substantially better in terms of C values, as shown in Table 2. Under the range studied, they did not observe activity decrease due to oversatura tion of the support, although they work at loads lower than 60 mg. of enzyme preparation/g d.w. chitin. C values for our immobilized lactase compare very favourably with those reported in the literature for a wide variety of supports[13].

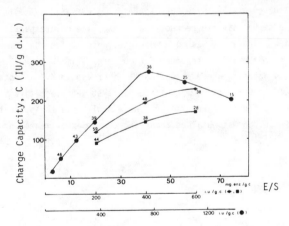

Fig.4. Charge capacity (C) of lactase as a function of enzyme load to chitin ratio, E/S (mg. enzyme preparation / g d.w.). This work (●); data of Stanley et.al at comparable conditions (■) and at t_1= o in proc. 4 (◆).
Numbers over experimental points represent corresponding immobilization yields (%).

Temperature and pH profiles were determined for both catalysts. Optimum temperature was 53° for invertase and 55° for lactase, pH optima were 4.3 and 4.0 which represent a displacement of about 0.3 units to the acid side, with respect to the soluble enzymes, as expected for chitin, which should have some unsubstituted amino groups bearing a possitive charge[34]. Kinetic studies were performed with immobilized invertase and lactase at their corresponding temperature and pH optima. Although substrate and product inhibition has been reported for invertase[26], our catalyst exhibited simple Michaelis-Menten kinetics, with no substrate nor product inhibition up to the levels of 200 g/l of sucrose and 50 g/l of glucose and fructose. Yeast and fungal lactases have been consistently reported to be competitively inhibited by galactose[2,10,12,16,36,54]. Glucose effect is seldom reported, although non competitive product inhibition[2], and activation[8] has been claimed. Our catalyst was inhibited by galactose in a total competitive mode. Glucose effect has been tested only on o - nitrophenyl galactopyranoside as substrate. Glucose, at 150 mM, produced an increase of 20% in Vmax, but also increased Km by 23%, indicating a complex pattern of modulation which should be studied in more detail.

Kinetic parameters for both immobilized enzymes are presented in Table 3 where, for comparison, the corresponding values for the soluble enzymes and for some catalysts reported in the literature are also presented, Diffusional restrictions were not significant for our catalysts as indicated by small differences in the Km values with respect to the soluble counterparts. In the case of lactase, the value of K_I was only moderately higher for the immobilized enzyme, differing from the values reported for commercial immobilized Sumilact (Sumitomo Chemicals, Japan)[46] where a substantial increase in K_I was obtained as a consequence of immobilization, making the catalyst less sensitive to galactose inhibition. External diffusional restrictions were negligible for our catalyst as a value of α = 0.086 was obtained according to the experimental procedure described by Pitcher[41].

241

Table 3. Kinetic parameters of immobilized invertase and lactase

ENZYME	Km	K_I	Vmax (μmol/min g)	T (°C)	Ref
INVERTASE					
-Chitin	68.5 (39.2)[*]		73.8	53	this work
- Bentonite	177 (143)		35	40	48
- Activated bentonite	133		75	40	48
- Amberlite IRC-50	18.7 (17.2)			40	37
- Activated clay	130 (140)	1110 (910)		50	26
LACTASE					
- Chitin (A.oryzae)	106 (89.8)	11.5 (8.0)	273	55	this work
- Phenol formaldehyde (A oryzae)	150 (100)	76 (6.4)	1000	30	46
- Polysulphone fibers (A. oryzae)	49 (46.9)	20	-	50	16
- Chitin (A.niger)	80 (80.)	-	53	45	21
- Chitin (A. niger)	85 (86)	-	105	40	45
- Porous glass (S. lactis)	19 (16)		257		
- Collagen (S. lactis)	22 (16)		308	30	54

* Values in parenthesis represent values for the corresponding soluble enzyme

Steady-state operation of a packed-bed reactor at 53°C, pH 4.3 with 10.7 g d.w. of immobilized invertase was performed at 75 g/l of sucrose and flow rates from 2 to 10ml/min. Results are presented in Figure 5 in terms of the effectiveness factor, defined as

$$\eta = \frac{\frac{si}{Km} \, x' - \ln(1-x')}{\frac{si}{Km} \, x - \ln(1-x)} \qquad (1)$$

where x is calculated from a model based on piston flow and simple Michaelis-Menten Kinetics

$$\frac{Vmax \; W}{F \quad Km} = \frac{si}{Km} \, x - \ln(1-x) \qquad (2)$$

Experimental results are in good agreement with the model at fluxes higher than 0.5 ml/min. cm^2

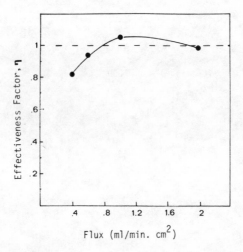

Fig.5 Steady-state performance of packed-bed reactor with immobilized invertase. Operating conditions are in the text.

A complete kinetic model for lactase is now being developed and steady-state reactor operations will be carried out afterwards.

Thermal stability for immobilized invertase and lactase was studied batchwise, in the absence of substrate at their corresponding temperature optima of 53 and 55°C. Half-lives were 25 and 10 days respectively, which is more than one order of magnitude higher than those of the corresponding soluble enzymes.

Long-term packed-bed reactor operation with immobilized invertase was carried out for 15 days at 53°C and pH 4.3 (results not shown), according to the scheme described for lactase. No enzyme was eluted from the reactor and conversion decreased only 5% after that period of time. Assuming steady-state operation and first-order thermal denaturation[50], half-life of the enzyme was estimated in 180 days.

Long-term packed-bed reactor operation with immobilized lactase was carried out for 15 days at 40°C and pH 4.0 using demineralized whey permeate (3.5 to 4% lactose) as substrate. The reactor was charged with 16 g d.w. of catalyst (4730 I.U.) and operated at a flow rate of 72 ml/h. Reactor was sanitized once a day with 0.1% acetic acid(a) and 0.05% alkylbenzyl-dimethyl ammonium chloride(d), for half an hour. No contamination was observed in the exit stream or inside the reactors, and no enzyme was eluted from the reactor during operation. Results are presented in figure 6. Initial conversion was 90% and after 15 days dropped only to 88.5%. Using the same procedure as for invertase, a half-life of 120 days was estimated for the catalyst, considering a kinetic model based only in competitive inhibition by galactose. The projected conversion at half-life was calculated to be 76%. Contrary to expected, reactor performance with USP lactose was somewhat lower with permeate, giving a projected conversion at half-life of 70%, although since that run we have improved our sanitization procedure. Actually, we did not expect significant differences in view of the very low protein and ion concentrations in our permeate, which have been reported to be the key factors affecting enzyme activity in whey[49].

An economic evaluation of our immobilized lactase should be done, taking as reference commercial immobilized fungal lactases available as those produced by Miles-Sumitomo, Röhm or British Charcoals & Mc. Donalds. In the meantime, other immobilization systems based on commercial lactases are being optimized to improve our results even further.

244

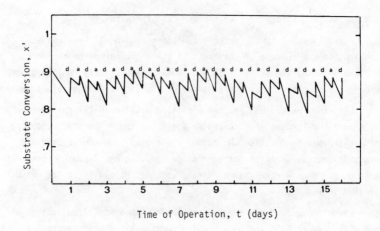

Time of Operation, t (days)

Fig.6. Long-term packed-bed reactor performance with immobilized lactase.
Operating conditions are in the text.

Acknowledgement. This work was supported by Grant N° 203 - 736 from the
the Division of Research, Universidad Católica de Valparaíso, Chile.

NOMENCLATURE

C	= charge capacity of support = EI/W	(IU/g d.w.)
d_r	= reactor diameter	(cm)
EI	= activity expressed in the immobilized catalyst	(IU)
EP	= unrecorvered activity	(IU)
ER	= residual activity in solution after immobilization	(IU)
F	= flow rate	(ml/min)
h	= reactor height	(cm)
h'	= mass transfer coefficient	(cm^3/sec)
K_I	= competitive inhibition constant	(mM)
K_m	= Michaelis constant	(mM)
si	= Inlet substrate concentration	(mM)
T	= Temperature	(°C)
t	= time of operation	(h)
t_1	= time of contact in stage 1 of immobilization	(h)
Vmax	= Maximum reaction velocity	(μmoles/min. g support)
$\bar{V}$max	= Maximum reaction velocity	(μmoles/sec.)
W	= Dry weight (d.w.) of support	(g)
x	= Theoretical degree of substrate conversion	
x'	= Experimental degree of substrate conversion = $\dfrac{si - s}{si}$	

$$Y = \text{Immobilization yield} = \frac{EI}{EI + ER + EP}$$

α	= Damkoehler number = $\bar{V}$max/h'Km
η	= Effectiveness factor = apparent reaction velocity/inherent reaction velocity

REFERENCES

1.   Anonymous. Drouin co-op scores a world first. Weekly Times, July 10 (1985).

2.   K. Cheng, J. Houng and A. Ling. Product inhibition of the enzymatic hydrolisis of lactose. Enzyme Microb. Technol. 7,510,514 (1985).

3.   Corning Technical Bulletin IBS-10B. Corning Glass Works, Corning, New York, U.S.A.

4.   G. Coton. The utilization of permeate from the ultrafiltration of whey and
      skim milk. Presented to the Dairy Federation, Geneva, Switzerland (1979).

5.   M. Daniels. Low-cost process for lactose hydrolysis with immobilized lactase.
      Food Technol. 39,68-70 (1985).

6.   M. Decleire, N. van Huynh, J. Motte and W. de Cat. Hydrolysis of whey by
      whole cells of Kluyveromices bulgaricus immobilized in calcium alginate
      gels and in hen egg white. Appl. Microbiol. Biotechnol. 22,438-441
      (1985).

7.   L. Dohan, J. Baret, S. Pain and P. Delalande. Lactose hydrolysis by
      immobilized lactase: semi-industrial experience, in Enzyme Engineering,
      H. Weetall and G. Royer (Eds.) Plenum Pres, Vol. 5, 279-283 (1980).

8.   E.D.C. Technical Bulletin 201 S-8/79. Enzyme Development Corporation, New
      York, New York, USA.

9.   O. Folin and H. Wu. A system of blood analysis. The J. of Biol. Chemistry
      38(1),81-110 (1919).

10.   V. Gekas and M. López-Leiva. Hydrolysis of lactose: a literature review.
      Process Biochem. 20(1), 2-12 (1985).

11.   C. Gertosio, J. Gutiérrez, T. Muzzio and A. Illanes. Evaluación técnico-eco
      nómica de una planta productora de suero de leche hidrolizado por vía
      enzimática. Anales del V Seminario Latinoamericano de Ciencia y Tecno-
      logía de Alimentos, p.39 Viña del Mar, Chile (1985).

12.   J. Giacin, J. Jakubowski, J. Leeder, S. Gilbert and D. Kleyn. Characterization
      of lactase immobilized on collagen: conversion of whey lactose by
      soluble and immobilized lactase. J. Food Science 39,751-754 (1974).

13.   N. Greenberg and R. Mahoney. Immobilization of lactase (B - galactosidase)
      for use in dairy processing: A review. Process Biochem. 16(2),2-8
      (1981).

14.   V. Holsinger. Lactose-modified milk and whey. Food Technol. 32,35-40 (1978).

15.   J. Hradil and F. Svec. Inversion of sucrose with invertase immobilized on
      bead DEAHP-cellulose: batch process. Enzyme Microb.Technol. 3,331-340
      (1981)

16.   L. Huffman and W. Harper. Lactose hydrolysis in batch and hollow fibre
      membrane reactors. New Zealand J. Dairy Sci. Technol. 20,57-63 (1985)

17.   L. Huffman and W. Harper. Beta-galactosidase retention by hollow fiber
      membranes. J. Dairy Sci. 65, 887-898 (1982).

18.   A. Illanes and Y. Gorgollón. Kinetics of extraction of invertase from
      autolysed bakers' yeast cells. Enzyme Microb. Technol. 7,510-514
      (1986).

19.   A. Illanes, R. Chamy and M. Zúñiga. Immobilization of invertase on

crosslinked chitin, in Chitin in Nature and Technology. R. Muzzarelli
et.al. (Eds.) Plenum Press (1985).

20. K. Imai, T. Shiomi, K. Sato and A. Fujishima. Preparation of immobilized
    invertase using poly (vinyl alcohol) membrane. Biotechnol. Bioeng.
    25,613-18 (1983).

21. T. Iwasaki, H. Hashiba and T. Kikuchi. Lactase immobilized on chitin with
    glutaraldehyde. T. Japan Kokai 7703892 (1977).

22. K. Kawashima and K. Umeda. Immobilization of invertase in poly acrylamide
    gels. Biotechnol. Bioeng. 16,609-614 (1974).

23. F. Knopf, M. Okos, D. Fouts and A. Syverson. Optimization of a lactose
    hydrolysis process. J. Food Science 44,896-900 (1979).

24. S. López-Covarrubias, A. Olano and M. Juárez. Obtención de B-lactosa, lactu
    losa y de mezclas de fructosa y galactosa a partir de suero de quesería
    desproteinizado. Rev. Agroquím. Tecnol. Aliment. 25, 355-361 (1985).

25. J. López-Santin, C. Sola, J. Paris and G. Caminal. Invertase immobilized on
    activated clay: properties and kinetic studies J. Chem. Techn.
    Biotechnol. 33B,101-106 (1985).

26. J. López-Santin, C. Solá and J. Lema. Substrate and product inhibition
    significance in the kinetics of sucrose hydrolysis by invertase.
    Biotechnol. Bioeng. 24,2721-2724 (1982).

27. O. Lowry, N. Rosenbrough, F. Lewis and R. Randall. Protein measurement with
    the Folin-phenol reagent. J. Biol. Chem. 19,265-275 (1951).

28. A. Manjón, F. Llorca, M. Bonete, J. Bastida and J. Iborra. Properties of
    B-galactosidase covalently immobilized to glycophase-coated porous-
    glass. Process Biochem. 20(1),17-22 (1985).

29. W. Marconi, F. Bartoli and F. Morisi. Improved whey treatment by
    immobilized lactase, in Enzyme Engineering, H. Weetall and G. Royer
    (Eds.) Plenum Press, Vol. 5, 269-278 (1980).

30. W. Marconi and Morisi. Industrial application of fiber-entrapped enzymes,
    in Applied Biochemistry and Bioengineering, L. Wingard et.al. (Eds.)
    Academic Press, Vol. 2,219-258 (1979).

31. W. Marconi, S. Gulinelli and F. Morisi. Properties and use of invertase
    entrapped in fibers. Biotechnol. Bioeng. 16,501-511 (1974).

32. M. Marek, O. Valentova and J. Kas. Invertase immobilized via its
    carbohydrale moiety. Biotechnol. Bioeng. 26,1223, 1226 (1984).

33. R. Mason and H. Weetall. Invertase covalently coupled to porous glass
    Biotechnol. Bioeng. 14,637,645 (1972).

34. R. Muzzarelli. Immobilization of enzymes on chitin and chitosan. Enzyme
    Microb. Technol. 2,177-184 (1980).

35. R. Muzzarelli, in Chitin. Pergamon Press, p. 90 (1977).

36. E. Okos and W. Harper. Activity and stability of B-galactosidase immobilized on porous glass. J. Food Science 39,88-93 (1974).

37. K. Ooshima, M. Sakimoto and Y. Harano. Characteristics of immobilized invertase. Biotechnol. Bioeng. 22,2155-2167 (1980).

38. M. Pastore and F. Morisi. Reduction of lactose in milk by entrapped B-galactosidase. IV. Results of long-term experiments with a pilot plant, in Enzyme Engineering, E. Pye and H. Weetall (Eds.) Plenum Press, Vol. 3, 537-542 (1978).

39. M. Pastore and F. Morisi. Lactose reduction of milk by fiber-entrapped B-galactosidase. Pilont plant experiments, in Methods in Enzymology XLIV, 882-830 (1976).

40. M. Pastore, F. Morisi and D. Zaccardelli. Reduction of lactose content of milk using entrapped B-galactosidase. Pilot plant studies, in Insolubilized Enzymes, M. Salmona et.al. (Eds.) Raven Press, 211-216 (1974).

41. W. Pitcher. Design and analysis of immobilized enzyme reactors, in Immobilized Enzymes for Industrial Reactors, R. Messing (Ed.) Academic Press, 151-199 (1975).

42. P. Raugel. Le marché des enzymes. Biofutur 20,29-34 (1984).

43. Röhm Technical Bulletin. V-850-1 Röhm Gmbh, Darmstadt, West Germany.

44. B. Sprössler and H. Plainer. Immobilized lactase for processing whey.Food Technol. 37,93-96 (1983).

45. W. Stanley, G. Waters, B. Chan and J. Mercer. Lactase and other enzymes bound to chitin with glutaraldehyde. Biotechnol. Bioeng. 17,315-326 (1975).

46. Sumitomo Technical Information Bulletin on Immobilized Sumylact. Sumitomo Chemicals, Japan.

47. J. Synowiecki, J. Sikorski and Z. Naczk. Immobilization of invertase on krill chitin with glutaraldehyde. Biotechnol. Bioeng. 17,315-326 (1975).

48. M. Tomar and K. Prabhu. Immobilization of cane invertase on bentomite. Enzyme Microb. Technol. 7,454-458 (1985).

49. E. van Gryethuysen, E. Flaschel and A. Renken. The influence of the ion content of whey on the pH-activity profile of the B-galactosidase from Aspergillus oryzae. J. Chem. Tech.Biotechnol. 35B,129-133 (1985).

50. W. Vieth, K. Venkatasubramanian, A. Constantinides and B. Dàvidson. Desing and analysis of immobilized-enzyme reactors, in Applied Biochemistry and Bioengineering. L. Wingard. et.al (Eds.) Academic Press, Vol. 1, 222-327 (1976).

51. L. Weckström, Y. Linko and P. Linko. Entrapment of whole cell yeast B-galactosidase in precipitated cellulose derivatives. Food Process Eng. 2,148-151 (1979).

52. A. Wiseman. Topics in Enzyme and Fermentation Technology Vol. 3, 267-294. Ellis Horwood Ltd. (1979).

53. A. Wiseman and J. Woodward. Immobilization of commercial bakers' yeast invertase. Process Biochem. 10,24-26 (1975).

54. J. Woychik, M. Wondolowsky and K. Dahl. Preparation and application of immobilized B-galactosidase of Saccharomyces lactis, in Immobilized Enzymes in Food and Microbial Processes, A. Olson and C. Cooney (Eds.) Plenum Press, 41-49 (1974)

CONTINUOUS ETHANOL PRODUCTION
WITH IMMOBILIZED YEAST CELLS
IN A PACKED BED REACTOR

S.V. Ramakrishna, V.P. Sreedharan and P. Prema
Regional Research Laboratory (CSIR),
Trivandrum 695019,
Kerala, India

Introduction

The conversion of plant carbohydrates into chemical feed stock has attracted the attention of scientists in recent years. Cassava (Manihot esculenta crantz) is one of the most efficient photosynthesising plants, having a carbohydrate content of 90% (dry basis). It mainly grows in tropical regions.

The cassava starch is considered to be inferior to maize starch, for use in the textile and paper industry because of lack of stability in viscosity. This is due to the weak intramolecular forces present in cassava. It is also sticky after cooking because of the longer chain length of the molecule and higher free amylose content; which has considerably hindered its utilization for dietery preparations. However, these characteristics have no bearing on hydrolysis and subsequent fermentation to ethanol.

Traditionally ethanol is produced from molasses, a sugar industry by-product, but of late, the molasses based distilleries are working under great constraints due to severe shortages of molasses. Under these circumstances the search for alternative sources for ethanol production became a necessity. The utilization of cassava starch for ethanol is well recognized[1] [2]

Ethanol is generally fermented in a batch process. It is now clear that the batch system suffers from many disadvantages and that the volumetric productivities are meagre[3](2.2g/l/h). The advantages of continuous fermentation with various systems such as cell recycle, vacuum fermentation etc. are well documented in the literature[4].

The recent concept of whole cell immobilization represents a novel approach to bioprocessing operations. A few immobilized cell systems have already been commercialized. This in turn has triggered a surge of research activity all over the world towards adopting this technology for various fermentation processes. It is well recognized that the concept of cell immobilization, obviates enzyme extraction/purification and can operate with higher stability and improved volumetric productivities[5] (50-100/l/h).

Immobilized cell systems for the production of ethanol have gained importance in recent years. Several authors have investigated immobilized cell reactor systems for ethanol production. Various inert supporting materials, different immobilization procedures and several substrates have been reported[6] [7] [9] [10]. For industrial use of immobilized cell columns, packed bed reactors are widely used because of their simplicity and better operational control. However, the main disadvantage of the packed bed reactor is compaction of the loading material due to liquid flow. These compaction effects are significantly greater when soft gel immobilized beads are employed. The bed is gradually compressed resulting in decreased liquid flow rate and hence productivity. Though some reports[11] have examined the compaction of immobilized beds, remedial measures for this have not been suggested. Recently Masakatsu Furui and Kiyokazu Yamshita[12] have proposed horizontal baffles in the column to reduce the compaction.

In the present study we have employed an immobilized yeast reactor for the continuous production of ethanol utilizing cassava starch hydrolysate. The incorporation of sieve plate baffles inside the reactor has gretly reduced the problem of compaction. The kinetics of yeast gel beads and reactor kinetics were studied. The reactor was found to be active over a period of 90 days.

Materials and methods

Microorganism and culture conditions:

Saccaromyces cerevisiae (NCIM 3288) was used in all experiments. The strain was grown in MYGP agar and stored at 4°C.

Cell propagation medium: Yeast extract 0.3%, pepton 0.5%, malt extract 0.5% and glucose 3%. The cells were cultured in a rotary shaker for 24 h and harvested by centrifugation.

Ethanol production medium: Yeast extract, 1.0% ammonium dihydrogen orthophosphate 0.1%, calcium chloride 0.07%, magnesium sulphate 0.01% and glucose 10%. pH was adjusted to 5.0.

Cassava starch hydrolysate:

Cassava starch was obtained by pulverising the commercially available cassava dry chips (sun dried) in a plate mill until the powder passed through a 40 mesh sieve.

Acid hydrolysis: A 20% slurry was prepared, mixed with 0.2 N HCL and subjected to pressure cooking at 15 psi for 2 hr. The cooked mash was neutralised with NaOH and filtered. The conversion was nearly total having 90% of the total reducing sugars as glucose[13].

Enzymatic hydrolysis: A 20% cassava starch slurry was prepared and heated up to

100 °C for 30 min: αamylase (EC 3.2.1.1) obtained from NOVO Industry was added at the required concentration and maintained at a temperature of 95 °C for 90 min. The mash was cooled to 60 °C and amyloglucosidase (EC 3.2.1.3) which was also obtained from NOVO was added at a pre-determined concentration after adjusting the pH to 4.5. The reaction was carried out for 120 minutes. The conversion was almost total, having 97% of the total sugars as glucose[13]. The hydrolysate was fed to the immobilized reactor after adjusting the pH to 4.5 and glucose concentration to 10%.

Immobilization procedure:

The principle of Ionatropic gelation of polyelectrolytes was employed for entrapment of the yeast. The harvested cells were mixed with 2-4% sodium alginate solution and extruded through a nozzle into a calcium chloride solution (1-2%). The geled droplets were cured for 3 h and stored at 4 °C. In a large scale preparation, the yeast-alginate slurry was passed through a multinozzle system. The size of the immobilized beads was 3.5-4.0 mm, and the bulk density was found to be 0.585 g/ml. The cell loading was determined by dissolving certain amount of beads in phosphate buffer (pH 7.0) and counting the cells, thus released, in a hematocytometer. The average cell loading was found to be 3.0 - 3.5 x 10^8/per bead.

Reactor:

A cylindrical glass column 50 mm i.d. and 600 mm height was used as a packed bed reactor. Provisions were made to take out the samples at different heights. For uniform distribution of the substrate, the bottom of the column was packed with glass beads of 4 mm diameter, which were supported on cotton. The column is fed from the bottom by a peristalic pump (Chemap AG). The product is withdrawn from the top with a similar pump operated at the same speed.

Sieve plate baffles:

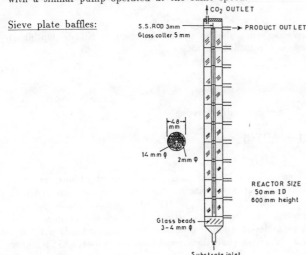

Fig. 1. Schematic Diagram of the Modified Reactor

Fig. 1 gives the schematic diagram of the reactor with baffles. An S.S rod 3 mm x 600 mm was fixed to a sieve plate which was placed over the glass beads at the bottom. A glass collar (5mm, i.d. and 50 mm length) was slid over the bottom perforated plate. Immobilized yeast beads were packed to a height of 50 mm and a second perforated plate was slid down over the bed; over which another collar of similar dimensions was placed. Again IY beads were packed to a height of 50 mm and over which a sieve plate was slid. The operation was repeated 10 times to get an immobilized yeast bed of 500 mm. At the top, a perforated plate was placed and the central S.S. rod was fixed by a rubber cork. The bed is virtually segmented in 10 equal parts with L/D ratios of 1. The baffles cannot move from their position due to the presence of a collar on both sides.

A similar type of packed bed reactor of larger capacity (100 mm i.d. and 1000 mm height) was also used in these studies.

Assay methods:

Reducing sugars were estimated by the Shaffer-Somogyi micromethod[14]. Glucose, concentration was estimated by an enzymatic method using glucose oxidase and peroxidase with o-diansidine. Composition of different sugars in cassava starch hydrolysate was estimated by HPLC (Waters Associates) on a Bondapak TM/carbohydrate analysis column using 85:15 acetonitrile: water at a flow rate of 2 ml/min with an R.I. detector. Alcohol was estimated by GLC (HP-5840 A gas chromatograph) on a porapak S column 6 ft in length using n-butanol as an internal standard at an isothermal temperature of 180 °C with a flame ionization detector.

Microscopic examination:

The immobilized cells were observed under an optical microscope (Nikkon Optiphot) and a scanning electron microscope (Jeol JSM-35C).

Results and discussion

Kinetics of immobilized yeast beads (IYB):

Considering the beads as biocatalysts, the classical kinetic theory of enzymes was applied. The substrate, pH and temperature optima for the beads were determined in a batch system by incubating known amounts of IYB in 100 ml of glucose medium for 48 hr. Alcohol was produced and left-out sugars were estimated for all the samples. The rate of reaction was calculated as moles of ethanol produced by 1 gm of beads per hour. The activity - pH profile (Fig.2) indicates that the immobilized yeast beads (IYB) have maximum activity at pH 5.0.

However, the difference in activity was barely influenced by the pH of the external medium[15] [16]. A temperature-activity profile

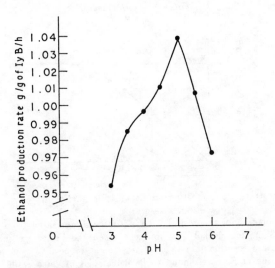

Fig. 2. pH - Activity Profile of IYB

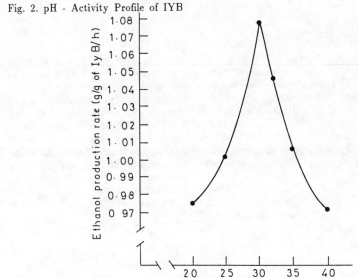

Fig. 3. Temperature - Activity

(Fig. 3) demonstrated that the IYB have the highest activity at 30°C. The substrate rate of reaction studies indicated that no substrate inhibition was noticed up to a 20%

glucose concentration (Fig. 4). However the conversion efficiency

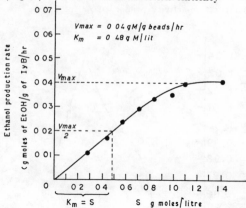

Fig. 4. Effect of Substrate Concentration on Ethanol Production Rate.

(experimental yield of ethanol/theoretical yield x 100) was decreased from 99% to 87.5%. A maximum of 89.3 mg/ml alcohol was produced when the IYB were subjected to 20% glucose. The maximum reaction rate (V_{max}) was found to be 0.04 g moles of ethanol/g of beads/h. The value of K_m was found to be 0.48 moles of glucose/100 ml.

Performance of packed bed reactor:

The small column (50 mm i.d. x 600 mm height volume 875 ml) was used for kinetic studies. In the initial experiments, the column was packed with IYB so as to have a 50 cm bed. The substrate was fed from the bottom and product was taken from the top. During the operation the bed was slowly lifted up due to liquid flow and CO_2 evolution, resulting in a decrease in the void volume. The compaction of the bed has hindered the reaction rates. However the effect of compaction was found to be significant in case of the bigger reactor (i.d. 100 mm, bed height 100 cm). Figure 5 shows the decrease in void volume and flow rate with respect to time. The void volume which was initially 0.33[17] (theoretical maximum for spherical bodies) was gradually decreased to 0.07 in 96 h, resulting in reduction in the flow rate of the substrate from 1.4 to 0.2 litres.

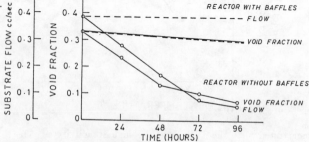

Fig. 5. Effect of Sieve Plate Baffle on Void Fraction of Liquid Flow.

257

This has seriously affected the reactor performance. The bed was lifted up to a height of 30 cm from the bottom. The decrease in void volume was found to be proportional to the distance from the bottom. The top 10 cm segment virtually became a thick block and IYB were completely distorted and flattened. To avoid the above situation, the column was modified with sieve plate baffles as described in the materials and methods. The baffled reactor of smaller size was used for kinetic experiments. Figure 6 gives the concentration of glucose and ethanol at different heights of the bed.

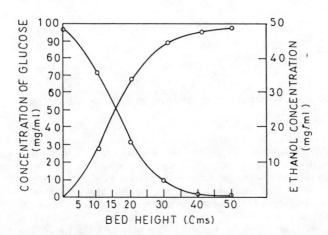

Fig. 6. Effect of Bed Height on Ethanol Production

It can be seen that about 90% conversion of glucose takes place within 30 cm height (L/D 6), when 10% glucose is fed through the reactor. Figures 7 and 8 show the volumetric productivities of the system as a function of the dilution rate.

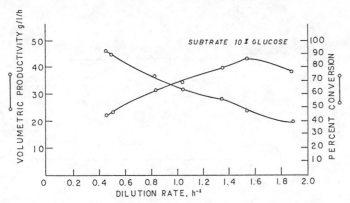

Fig. 7. Effect of Dilution Rate on Productivity and Glucose Conversion.

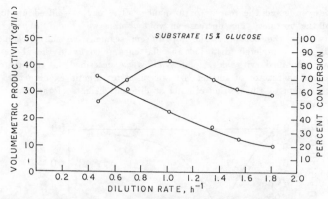

Fig. 8. Effect of Dilution Rate on Productivity and Glucose Conversion.

When 10% glucose was fed as a substrate, at a low dilution rate $(0.46\ h^{-1})$ the conversion of glucose to ethanol is 92.8%, but the volumetric productivity is only 22g/L/h. As the dilution rate is increased gradually from 0.46 to 1.88, the glucose to ethanol coversion rate has fallen steeply from 92.8% to 39%. However, the volumetric productivities increased with the increase in dilution rate to $1.54\ h^{-1}$ and then slightly lowered thereafter. The maximum volumetric productivity of 42% g/L/h was obtained at a dilution rate of $1.54\ h^{-1}$ corresponding to a conversion rate of 47.8%. For a glucose feed of 15% similar results were obtained. The maximum volumetric productivity ws 41.89 g/L/h at a dilution rate of $1.15\ h^{-1}$ corresponding to a coversion rate of 45.9%. This has indicated the need to operate the rector at a higher dilution rate. Based on these results the larger reactor was also modified with baffles and operated with cassava starch hydrolysate (glucose level adjusted to 10%). The reactor was working for more than 90 days. The conversion rate of commercially available cassava hydrolysate to ethanol was in the range of 70-80% (Fig. 9).

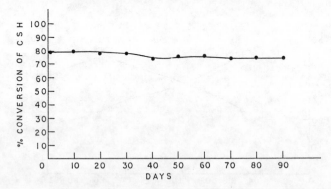

Fig. 9. Reactor Performance with Commercially Available CSH

The lower conversion rates are found to be due to the presence of maltose and other oligosaccharides in the hydrolysate. A separate study was carried out to find out the influence of the composition of sugars in cassava starch hydrolysate on alcohol production[18]. It was found that the strain used was capable of utilizing either glucose or maltose when they were present alone, but in a mixture it utilises glucose preferentially. In a continuous fermentation, the effect was significant as it was seen in the reactor. In subsequent experiments, the hydrolysate was prepared with a view to contain at least 90% of the total sugar present as glucose.

Microscopic examination of the IYB:

Figure 10 shows the SEM photographs of an IYB surface and cross section.

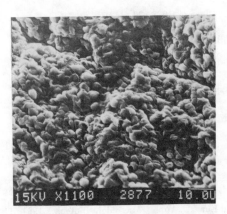 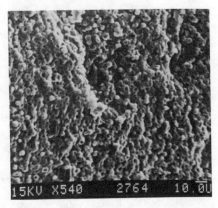

Fig. 10. Scanning Electron Micrographs

The IYB surface is covered with micropores 2-3 mm in diameter through which the transport of substrate and product is taking place. The pores facilitate the passage of CO_2 which was a co-product in the reaction[19].

Conclusions:

Our study has shown that the immobilized yeast cells can be successfully used for continuous ethanol production from cassava starch hydrolysate. A modified reactor with internal sieve plate baffles helped in reducing the problem of compaction considerably. A maximum productivity of 42 g/L/h can be achieved with the present system. The Cassava starch hydrolysate should contain at least 90% of the total sugars as glucose for efficient conversion. It is possible to run the reactor for 3 months continuously.

Acknowledgements:

The authors acknowledge the help rendered by Dr J. Madhusudhana Rao, V.B. Manilal, Rachel Samuel, T. Sita Ratnakumari, S. Sisupalan and Kamalabai Amma. They are also grateful for the encouragement and support given by Dr C.S. Narayanan, Head, Food Division, Dr A.G. Mathew, Head, Chemical Group and Dr A.D. Damodaran, Director of the laboratory.

References;

1. Poulsen, P.B. and Garberg, P. Symp. Proceedings of AHARA-82, 1982.

2. Ramakrishnan, S.V., Raja, K.C.M. Emilia Abraham, T. and Sreedharan, V.P. Proceedings of Bio-Energy Society, First Convention & Symposium, 1985, 101.

3. Cysenski, G. and Wilke, C.R. Biotech. Bioeng., 1978, 20, 1421.

4. Maiorella, B.L., Blanch, H.W. and Wilke, C.R. Biotech. Bioeng., 1984, 26, 1003.

5. Kolot, F.B. Proc. Biochem., 1984, 2, 7.

6. Chibata, I. in Immobilized microbial cells, ACS Symposium series 1979, 106, 187.

7. Ghose, T.K. and Bandyopadhayay K.K. Biotech. Bioeng., 1982, 24, 794.

8. Nagashima, M., Azuma, M., Noguchi, S., Inuzuka K. and Samejima, H., Biotech. Bioeng., 1984, 26, 992.

9. Nojima, S., Chem. Eco. & Eng. Review, 1983, 15, 4,17.

10. Wada, M., Kato, J., and Chibata, I., Eur. J. Appl. Microbiol. Biotech., 1981, 11, 67.

11. Furasaki, S., Okamura, Y., Miyauchi, T., J. Chem. Eng., Japan, 1982, 15, 148.

12. Furui, M. and Yamashita, K., J. Ferment. Technol., 1985, 631, 73.

13. Emilia Abraham, Krishnaswamy, C., & Ramakrishna, S.V. (unpublished work).

14. Shaffer, P.A. and Somogyi, J. Biol. Chem., 1933, 100, 695.

15. Cho. G.H. Choi, C. Dochoi, Y., and Han M.H.J. Chem. Tech. Biotechnol, 1982, 32, 959.

16. Williams, D., Munnecke, Douglas, M., Biotech. Bioeng., 1981, 23, 1819.

17. Perry's Chemical Engineer's Hand Book. Ed. D.W. Green 6th Edition, 1984.

18. Prema, P., Ramakrishna, S.V. and Madusudhana Rao, J., (In press).

19. Sreedharan, V.p., and Ramakrishna, S.V. (In press).

IMMOBILIZED BIOCATALYSTS AND ANTIBIOTIC PRODUCTION: BIOCHEMICAL, GENETICAL AND BIOTECHNICAL ASPECTS

E.J. Vandamme
Laboratory of General and Industrial Microbiology
State University of Ghent
Coupure Links 652,
B-9000 Ghent, Belgium

ABSTRACT

Immobilized biocatalyst technology for producing known or new antibiotics is gaining much interest. An evaluation is presented of the applicability of this concept in the fascinating field of bioconversion and fermentation.

Use of immobilized enzymes, organelles and viable cells to synthesize antibiotics as an alternative to conventional fermentation is discussed. In vitro total enzymatic synthesis is illustrated with the "multi-enzyme thio-template mechanism" (gramicidin S, enniatin) and with the enzymes involved in penicillin and cephalosporin formation. Total synthesis of peptide antibiotics, based on immobilized viable cells has been demonstrated with penicillin, bacitracin, nisin, patulin, tylosin, cyclosporine,... and a few other antibiotic compounds.

As an industrial example of the use of enzymes or cells to convert peptide antibiotics into therapeutically useful derivatives, free and immobilized penicillin acylases, producing the penicillin nucleus 6-aminopenicillanic acid (6-APA), are reviewed as well as their potential to synthesize semi-synthetic β-lactam antibiotics (penicillins, cephalosporins, nocardicins, monobactams). Microbial acylases acting on cephalasporinand yielding valuable intermediary (7-ACA and 7-ADCA) or end products have also gained industrial interest. Stereospecific enzymic side chain preparation for semi-synthetic penicillin and cephalosporin production has also recently reached the industrial stage.

261

Bioconversion possibilities with novel peptide β-lactam compounds, rifamycins and aminoglycosides are reviewed. In each case, biochemical, genetic and biotechnical aspects are discussed.

These examples of simple single-step (bioconversion) as well as complex multi-step enzyme reations (fermentation) point to the vast potential of immobilized biocatalyst-technology in fermentation science, in organic synthesis and in biotechnological process in general.

INTRODUCTION

Of the more than 6,000 known natural microbial compounds which display antibiotic activity, about 150 are produced on a large scale and find use in medicine and agriculture, mainly as antibacterial, antifungal or antiviral agents. A minority find use as antitumour agents, immune modulators, as anthelminthical, herbicidal, antiprotozoal, piscicidal or anticoccidial agents, feed additives, food preservatives or plant disease controllers (Demain, 1983; Vandamme, 1984). (Scheme 1).

ANTIBIOTIC COMPOUNDS

ANTIBACTERIALS	ANTITUMORS
ANTIFUNGALS	ANTIVIRALS
COCCIDOSTATS	FEED SUPPLEMENTS
ANTIPROTOZOALS	FEED SUPPLEMENTS
PLANT-DISEASE CONTROLLERS	PISCICIDALS
ANTHELMINTICS	HERBICIDALS
IMMUNO-MODULATORS	

All these commercial antibiotics (except chloramphenicol and cellocidin) are produced by micro-organisms in conventional fermentation processes. In a few cases, the natural microbial product can be chemically or enzymatically converted into a so-called semisynthetic antibiotic with superior therapeutic properties (Vandamme, 1980; Sebek, 1980). In the last decade, several attempts have been made to apply the immobilized enzyme or immobilized-cell principle to antibiotic fermentation and biconversion processes (Klibanov, 1983; Vandamme, 1983, 1984). Presently, several antibiotic bioconversions involving single-step reactions are run industrially with immobilized enzymes or cells as biocatalysts.

263

The replacement of conventional fermentation (which usually involves a complex-multi-step reaction sequence) with immobilized enzymes (Demain and Wang, 1976; Madry et al., 1984; Jensen et al., 1984; Wolfe et al., 1984) or immobilized viable cell technology (Demain and Wang, 1976; Egorov et al., 1978; Venkatasubramanian and Vieth, 1979; Morikawa et al., 1979a,b; 1980; Freeman and Ahronowitz, 1981; Veelken and Pape, 1982; Deo and Gaucher, 1983; Foster et al., 1983) also seems to offer great potential (Scheme 2).

<u>NOVEL PRODUCTION OF ANTIBIOTICS</u>

- MULTI-STEP ENZYMIC TOTAL SYNTHESIS

- IMMOBILISED VIABLE CELL-FERMENTATION ⟷ CONVENTIONAL FERMENTATION

- BIOCONVERSION WITH IMMOBILISED BIOCATALYSTS

<u>Acellular Total Biosynthesis of Antibiotics</u>

As mentioned above, only 2 out of the about 150 commercial antibiotics are presently made by chemical synthesis, despite the fact that chemical routes are known for many important antibiotics. The economics of multi-reaction chemical synthesis of such complex biological molecules are simply too unfavourable. As organic synthesis is virtually impossible, antibiotic synthesis by traditional fermentation also has its drawbacks: as secondary metabolites, their synthesis is usually delayed until growth declines; antibiotic "synthetases" are rapidly inactivated; substrate conversion into antibiotic is inefficient and strain degeneration presents a major problem.

To overcome these problems, attempts have been made to replace fermentation or cellular synthesis by acellular processes, i.e. total enzymic synthesis <u>in vitro.</u> In such a process, it is the ultimate aim to use isolated, stabilized and immobilized enzymes, which, in sequential reactions, perform the total synthesis of an antibiotic upon addition of its precursors, ATP and cofactors.

The system that is best understood is the biosynthesis of gramicidin S (GS), a cyclic decapeptide antibiotic produced by certain <u>Bacillus</u> <u>brevis</u> strains (Figure 1). The number of enzymes in the biosynthetic pathway is small, i.e. GS synthetase 1 and 2.

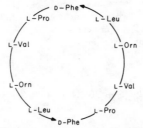

Figure 1 Structure and schematic representation of gramicidin S

264

The enzymes, substrates and cofactors involved have been characterized and the biosynthetic mechanism has been resolved in detail. Both enzymes have been purified to homogenity. Phosphopantetheine is the cofactor (Figure 2) (Kleinkauf and Koischwitz, 1980; Vandamme, 1981; Krause et al., 1985).

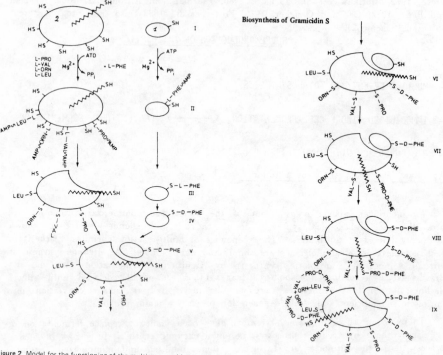

Figure 2 Model for the functioning of the multienzyme thiotemplate mechanism in gramicidin S biosynthesis (1 = GS synthetase 1, 2 = GS synthetase 2).[14] I, Schematic representation of GS synthetases 1 and 2. The zig-zag line represents 4'-phosphopantetheine. II, Activation of amino acids and binding on GS synthetases 1 and 2. III, Thioester binding of amino acids and conformational change of GS synthetase 2. IV, Racemization of L-Phe to its D-isomer. V, GS synthetases 1 and 2 complex, loaded with thioester-bound amino acids. VI, Transfer of D-phenylalanine on GS synthetase 1 to a thiol site of GS synthetase 2. VII, Formation of the first intermediate dipeptide. VIII, Transfer of the dipeptide to the 4'-phosphopantetheine arm. IX, Two pentapeptides cyclize 'head to tail' in an intramolecular reaction on GS synthetase 2 [Reproduced from Vandamme, E. J. in *Topics in Enzyme and Fermentation Biotechnology* (Wiseman, A., ed.) Ellis Horwood, Chichester, 1981, vol. 5, p. 185, by permission of Ellis Horwood Limited]

The mechanism involved has been named "multi-enzyme thiotemplate" biosynthesis, and it requires, in addition to the two enzymes, only the building block amino acids or their analogues as precursors, ATP as an energy source, and Mg^{2+} ions. The overall biosynthesis scheme can be simplified as follows:

$$\begin{matrix} 2\ Phe \\ 2\ Pro \\ 2\ Val \\ 3\ Orn \\ 2\ Leu \end{matrix} + 10\ ATP \xrightarrow{\text{synthetase 1 and 2}} 1\ GS + 10\ AMP + 10\ PP_i$$

Using the two GS synthetases, isolated from high-yield gramicidin S-producing B. brevis ATCC 9999 cells, gram quantities of gramicidin S have been produced in vitro by the total enzymic method (Demain and Wang, 1976; Demain et al., 1976).

The transient formation of the GS synthetases during growth of B. brevis makes it difficult to harvest large quantities of these labile enzymes. Recent experiments have confirmed the presence of particulate GS synthesases, which might be easier to recover and to stabilize outside the cells (Poirier and Demain, 1981). The two GS synthetases could be coimmobilized to construct a bioreactor, coupled to an enzymic ATP-regeneration system which may allow for a continuous process to be developed. Furthermore, instead of using expensive amino acids as substrates, cheap protein hydrolysates should be tested.

The depsipeptide antibiotic enniatin (Figure 3) produced by the fungus Fusarium oxysporum, is also produced via the thiotemplate mechanism using here one multifunctional enzyme, enniatin synthetase (M.W. 250,000). This partially purified enzyme was immobilized by adsorption to propylagarose, with 45% retention of original activity. Selective synthesis of several emiatin homologues was achieved, including non-natural depsipeptides (Maudry et al., 1984).

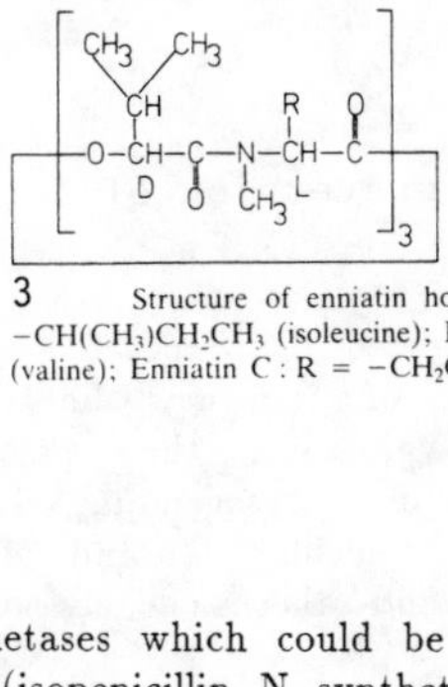

Fig. 3 Structure of enniatin homologues. Enniatin A : R = −CH(CH₃)CH₂CH₃ (isoleucine); Enniatin B : R = −CH(CH₃)₂ (valine); Enniatin C : R = −CH₂CH(CH₃)₂ (leucine)

Other antibiotic synthetases which could be exploited in total enzymatic synthesis processes, include cyclase (isopenicillin N synthetase), epimerase, ring expansion enzyme (expandase) and oxidizing enzymes, which are important in penicillin and cephalosporin β-lactam biosynthesis (Jensen et al.,1982; Leungo et al.,1986); these enzymes seem to be membrane bound in fungi (Penicillium and Acremonium), but appear to be soluble in Streptomyces clavuligerus (Figure 4).

Jensen et al., (1984) reported on the immobilization in a DEAE-tris-acrylgel of the above mentioned four soluble enzymes from Streptomyces clavuligerus they demonstrated continuous multistep conversion of the natural ρ-(L-α-amino adipyl) -L-cycteinyl-D-valine (ACV)-tripeptide into B-lactam antibiotics (Wolfe et al., 1984).

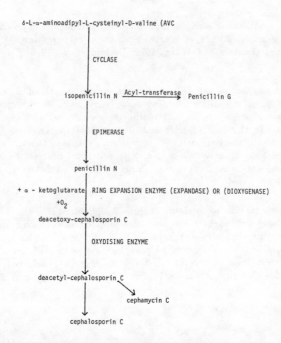

δ-L-α-aminoadipyl-L-cysteinyl-D-valine (AVC

CYCLASE

isopenicillin N →(Acyl-transferase)→ Penicillin G

EPIMERASE

penicillin N

+ α - ketoglutarate RING EXPANSION ENZYME (EXPANDASE) OR (DIOXYGENASE)
+0$_2$

deacetoxy-cephalosporin C

OXYDISING ENZYME

deacetyl-cephalosporin C

cephamycin C

cephalosporin C

Fig. 4.   Enzymes involved in conversion of ACV into penicillins
           and cephalosporins.

Recently, Luengo et al., (1986) reported the direct one-step enzymatic synthesis of penicillin G by in vitro cyclization of the new linear tripeptide-like molecule, phenylacetyl-L-cysteinyl -D-valine (PCV), using the partially purified cyclase of Penicillium chrysogenum and Acremonium chrysogenum. This finding may have an important industrial relevance, since penicillin G and derivatives with improved properties, may be produced from PCV and other analogues using immobilized cyclases.

Membrane or organelle-localized enzyme complexes have also been found to be able to catalyse partial or even total antibiotic synthesis.

Kurzatowsky et al.(1982) isolated vesicles from P. chrysogenum PQ-96 protoplasts which displayed penicillin G total biosynthesis capacity.

Immobilization of these vesicles in a calcium alginate gel resulted in 44% of the activity of native vesicles. This activity remained stable after 240 h storage at 4° C, whereas, the native vesicles had lost all biosynthetic activity within 60 h. The use of such a stablizied superior organelle-biocatalyst in a continuous bioreactor might provide for improved penicillin production.

Potential applications of total emzymic synthesis of antibodies are:

-efficient production of currently used and known antibiotics.

-the production of biosynthetic intermediates as starting materials for chemical conversion or bioconversion into new antibiotics by omitting certain enzymes from such systems.

-total enzymic synthesis of novel antibiotics since the in vitro specificities of antibiotic-forming enzymes are not very strict; substrates could be used which could otherwise not be incorporated by whole cells because of permeability constraints, degradation or toxicity.

-the synthesis of other complex molecules of medical, nutritional, agricultural or industrial importance.

Immobilized Viable Cells for Antiobiotic Fermentation

The problems encountered as yet with complex total enzymic synthesis processes using cell-free (immobilized) whole viable cells (Vandamme, 1976; Chibata and Tosa, 1981; Fukui and Tanaka, 1982). This concept would be particularly valuable for those antibiotics which are, or can be, excreted into the fermentation broth. Fermentation of penicillin G, cephamycin, thienamycin, candicidin, misin, bacitracin, colistin, tylosin, nikkomycin, patulin, cyclosporin A and a few other compounds has already been tested in this respect (Veeklken and Pape, 1982; Deo and Gaucher 1983; Vandamme, 1983; Foster et al., 1984, Karube et al., 1984, et al., 1984).

Immobilized viable cell systems can offer several important advantages over conventional fermentation, such as the possibility of continuous operation (at high dilution rates with no risk of washout) or plug flow mode of action, column fermentation, reduction of non-productive growth phases; faster reaction rates become possible at increased cell density, higher yields can be achieved, together with easier rheological control and control of slow cell reproduction.

Details on the use of immobilized fungal Penicillium chrysogenum mycelium to produce penicillin G have been presented by Morikawa et al., (1979a); Wang et al., (1984) and by Karube et al., (1984). Production of patulin by immobilized Penicillium urticae mycelium was reported by Deo and Gaucher (1983). The fungal mycelium of Tolupocladium inflatum has been immobilized in carrageenan-beads and tested in an airlift bioreactor for cyclosporin A (cyclosporine) (Figure 5) production; this antifungal cyclic undecapeptide antibiotic also displays important immunosuppressive properties in protecting allo- and xenografts from rejection and has found wide use in transplant-surgery (Green,1981). The cyclosporin A recovered from the bio-reactor by a single-step solvent extraction was essentially free of contaminating media and endogenously produced metabolites (Foster et al., 1983).

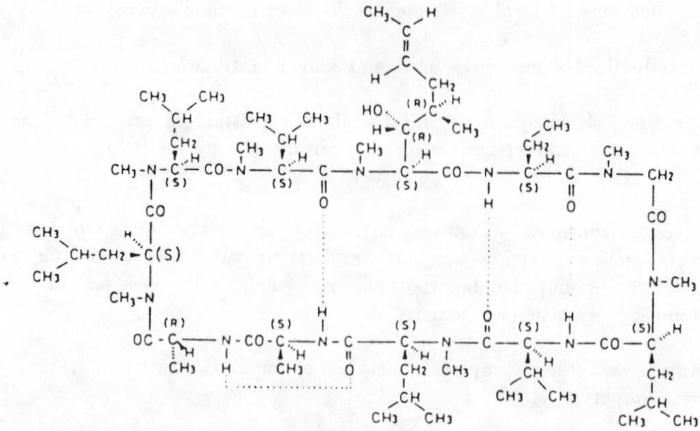

268

Fig. 5 Structure of Cyclosporin A

Immobilization of fungal cells also avoids problems associated with pellet-morphology and broth rheology; it provides the fermentation liquid with desirable hydrodynamic properties (lower viscosity and foaming, lower power consumption and aeration rates.

Gel-entrapment procedures are based on polyacrylamide (PAA).

To avoid inactivation of the antibiotic synthetases in viable cells by PAA-monomers and by polymerisation-heat, prepolymerised PAA-chains were used to immobilize Streptomyces clavuligerus cells followed by cross-linking (Freeman and Aharonowitz, 1981) ; thus immobilized cells, harvested from the early growth phase, produced B-lactam antibiotics in yields comparable to those of free resting cells.

Streptomyces T-59-235 cells, immobilized in Ca-alginate gel produced the macrolide tylosin (Figure 6) over a prolonged period; the nucleoside peptide antibiotic nikkomycin could be produced with Ca-alginate immobilized Streptomyces tendae; relative productivity was 40 - 50% as compared to that of free myceliem (Veelken and Pape, 1982); in a continuous air bubble reactor, nikkomycin production was maintained up to 640 hours with maximal specific productivity comparable to a batch fermentation process.

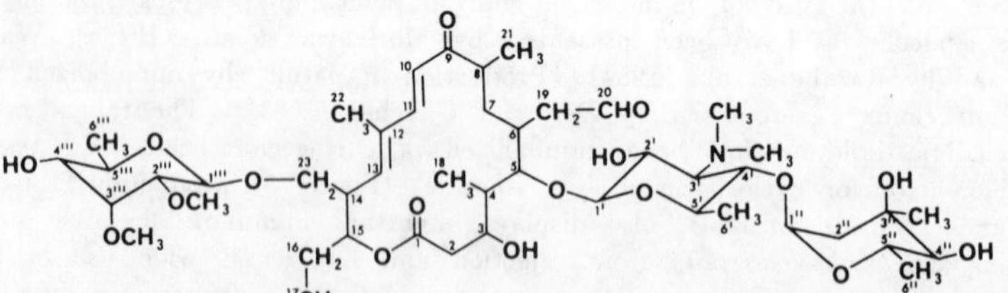

Fig. 6 Structure of tylosin

Bacillus cells were immobilized in PAA-gel and used to produce continuously bacitracin (Morikawa et al., 1980); reactivation of the cells by successive addition of nutrients and periodic washing of the gel to remove bead-surface growth resulted in a gradual increase in activity. This system allowed to produce antibiotic from a simple nutrient (peptone), and provided an easy antibiotic recovery from a cell-free liquid.

Nisin production by polyacrylamide gel-immobilized Streptococcus lactis cells has been reported by Ergorov et al., (1978). Nisin is an economically important polypeptide antibiotic compound. It is active against Gram-positive organisms, including related streptocci. Nisin resembles a bacteriocin rather than an antibiotic. Among antibiotics, it has the unique function of being used as a biological food preservative. It is non-toxic and, being a polypeptide, any residues remaining in food are digested (Hurst, 1981).

Whether nisin biosynthesis is coded for exclusively by the pSN-plasmid, is still a matter of debate. If so, this would open up a new field for such bacteriocin production in general (prevalent among lactic acid bacteria) via recombinant DNA- -technology (GASSON, 1984).

The above systems are typical examples of synthesis by fixed viable cells of secondary metabolites, normally non-growth associated complex fermentation products. The physiological state, culture age and viability of the cells in the immobilized reactor are of prime importance for effecting such complex multi- enzyme reactions.

Immobilised Biocatalysts and Antiobiotic Bioconversion

A) Penicillin Bioconversions

The worldwide demand for the semisynthetic penicillins has brought the penicillin nucleus, 6-APA into a central position as a major pharmaceutical product (Figure 7). As many as 16 different semisynthetic penicillins, all derived from 6-APA, are in widespread clinical use today. Bulk 6-APA production can now be achieved either by chemical or enzymic hydrolysis. However, with the introduction of methods of immobilization of enzymes or cells that promote stability and prolonged high activity that can result in reuse and continuous bioconversion, and with high energy prices, the enzymic splitting currently appears to display better economics. The microbial enzymes that hydrolyse penicillins into 6-APA or acylate 6-APA have been named penicillin acylases (EC. 3.5.1.11).

Recently, the name A-acylamino-B-lactam hydrolase has been proposed (Ryu and Nam, 1985). The enzyme occurs in a wide range of bacteria, actinomycetes, yeasts and fungi (Vandamme, 1980; Lowe, 1985).

Highly productive and genetic engineered strains are available for industrial bioconversion of penicillin V or G into 6-APA (Vandamme, 1985; Schomer et al., 1984; Olsson et al., 1985). Fermentation processes for penicillin acylases as well as the properties of these enzymes, have been reviewed (Vandamme, 1981; Ryu and Nam, 1983; Kasche et al., 1984; Mahajan and Borkar, 1984; Daumy et al., 1985). Several types of penicillin acylase can be recognized depending on their substrate spectrum: penicillin V-acylase, penicillin G-acylase and ampicillin acylase (Figure 7), (Scheme 3).

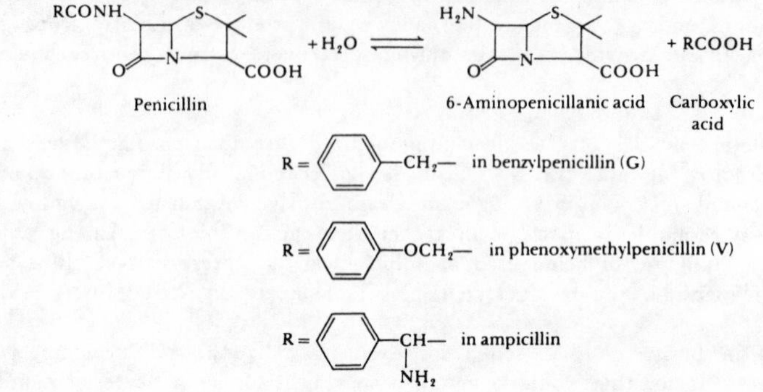

Fig. 7 Formation and Reacylation of 6-Aminopenicillanic
Acid (6-APA) by Penicillin Acylases.

INDUSTRIAL ACYLASE PRODUCTION

1) PENICILLIN G ACYLASE (E.C. 3.5.1.11)

BAYER, W. G.
BEECHAM, U. K.
GIST-BROCADES, NL
SNAM PROGETTI, IT   4000 T of 6-APA/YEAR
PFIZER, USA
TOYO YOZO, JAPAN

2) PENICILLIN V ACYLASE

BIOCHEMIE, AUSTRIA
NOVO INDUSTRI, DK   500 T of 6-APA/YEAR

3) CEPHALOSPORIN C ACYLASE : ?

4) GLUTARYL-7-ACA ACYLASE
+ D-AMINO ACID OXIDASE (E.C. 1.4.3.3)

ASAHI CHEMICAL IND. CO., JAPAN   7-ACA

5) HYDANTOINASE (E.C.3.5.22) : AGROBACTERIUM, BACILLUS

KANEGAFUCHI, JAPAN, SINGAPORE
RECORDATI, (SNAM PROGETTI), IT   50 TON D-PHENYL-CLYCINE/YEAR

The industrial enzymic route for 6-APA production was developed around 1960. Initially, cell suspensions of active strains could only be used once and the productivity was very low, about 0.5 to 1 kg of 6-APA per kg E. coli suspension. Novel methods to immobilize E. coli cells increased productivity up to 50 kg of 6-APA per kg of immobilized catalyst. Bayer applied a poly(methacryl)glutaraldehyde method, while Pfizer used binding to glucidyl methacrylate polymers. Beecham used a cyanogen bromide-activated dextran/Sephadex method and Astra a cyanogen bromide-activated Sephadex method. Current productivities are within the range 100-1000 kg 6-APA/kg; however, few data have been published by the companies involved. It can only be assumed that 60% of all 6-APA produced today is made by the immobilized catalyst route. This means that about 4000 ton 6-APA is produced by 15-30 ton immobilized penicillin acylase catalyst Poulsen,1984). Industrial 6-APA producers include Bayer, Gist-Brocades, Beecham, Glaxo, Astra, Spofa, Hoescht, Cipan, Biochemie, Antibioticos and Snam Progetti in Europe; Pfizer, Bristol-Myers, Squibb, Wyeth in the USA; Kyowa Hakko Kogyo and Toyo Yozo (Japan) and Ranbaxy (India) and Yung Jin Pharmaceuticals (S. Korea) in Asia (Scheme 4).

INDUSTRIAL PRODUCTION OF 6 - APA

ANTIBIOTICOS, SPAIN	KYOWA HAKKO, CO., JAPAN
ASAHI CHEM., JAPAN	NOVO INDUSTRI, DK
ASTRA, SWEDEN	RANBAXY, INDIA
BAYER, W. G.	ROHM, W. G.
BEECHAM, U. K.	SNAM-PROGETTI, IT
BIOCHEMIE, AUSTRIA	SPOFA, CZECHOSLOVAKIA
BRISTOL-MYERS, USA	SQUIBB, USA
CIPAN, PORTUGAL	TOYO ZOZO, JAPAN
GIST-BROCADES, THE NETHERLANDS (ALSO CHEMICALLY)	WYETH, USA
GLAXO, U.K.	YUNG-JIN, S-KOREA
HOECHST, W. G.	COMPANIES IN CHINA, USSR

Numerous immobilization techniques for penicillin acylase enzyme or cells have been tried out. Adsorption, crosslinking, covalent and physical attachment and entrapment methods have been used. Both soluble and insoluble carriers have been tested. A comprehensive survey of these aspects has been compiled recently Vandamme, 1980) (Scheme 5 and 6).

Industrial application of immobilized penicillin acylases is still confronted with problems (Savidge, 1984; Harrison and Gibson, 1984). Upon penicillin hydrolysis into 6-APA at pH 7.0 - 8.0 at 37° C, the side-chain acid is liberated, which causes a drop in the pH and this pH change results in a slower reaction rate. A higher starting pH is not desired because of B-lactam ring hydrolysis and penicillin inactivation. A strict pH control is thus necessary during this bioconversion process, which is therefore difficult to run in a continuous packed-bed reactor (Figure 8).

TYPICAL PROCESS DATA FOR IMMOBILISED PENICILLIN G ACYLASE

ENZYME PRODUCT	RIGID GRANULES OR DEXTRAN/SEPHADEX: E. COLI ACYLASE	ACYLASE (B. MEGAT ERIUM) COVALENT BOND TO GLUTARALDEHYDE-ACTIVATED POROUS POLY-ACRYLONITRILE FIBRES
REACTOR	COLUMN	PARALLEL COLUMN SYSTEM
SUBSTRATE	PENICILLIN G OR CERTAIN SEMISYNTHETIC CEPHALOSPORINS (4 - 15% W/V)	PENICILLIN G (10% W/V)
TEMPERATURE	35-40°C	30-36°C
INLET pH	7-8	8.4 ± 0.1
OPERATING LIFE	2000 - 4000 HOURS	APPROX. 1200 HOURS
PRODUCTIVITY	1000 - 2000 KG/KG ENZYME	500 - 700 KG/KG ENZYME
	ACCORDING TO BEECHAM, U.K.	ACCORDING TO TOYO YOZO, JAPAN

PROCESS DATA FOR IMMOBILISED PENICILLIN V ACYLASE

ENZYME PRODUCT	PENICILLIN V ACYLASE CELL HOMOGENATES COVALENTLY BOUND WITH GLUTARALDEHYDE
REACTOR	STIRRED BATCH OR COLUMN
SUBSTRATE	PENICILLIN V (4-12% W/V)
TEMPERATURE	35°C
INLET pH	7.5 - 8.0
OPERATING LIFE	1000 - 2000 HOURS
PRODUCTIVITY	200 - 600 KG/KG ENZYME
	DATA FROM NOVO INDUSTRI, DENMARK

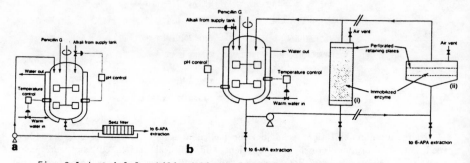

Fig. 8 Industrial Immobilized Biocatalyst Reactors for 6-Amino-penicillanic Acid (6-APA) Production.(a) Stirred Tank Reactor with External Recovery of Immobilized Enzyme. (b) Recirculation Reactor:(i) High Aspect Ratio Column; (ii) Low Aspect Ratio Column(Reproduced by Kind Permission of T.A. Savidge).

Internal in situ pH-regulation by enzymatic action has been achieved by co-immobilization of urease and adding urea in the inlet stream (Rousseau et al., 1984). Ammonia produced by urease action neutralizes the acid formed by side chain splitting of penicillin. Penicillin acylase has a more alkaline pH-optimum (pH 7.5 - 8.0) than urease. Thus as the pH rises, the activity of the acylase (acid forming enzyme) increases relatively to the acid consuming (urease) enzyme; this results in a pH-drop (and vice versa). This is a self-regulating system!

In industry, a batch reactor is generally used. Such a process is then faced with another aspect of the catalytic properties of the enzyme, namely end-product inhibition. So, in this particular case, it is in fact the nature of the biochemical reaction that determines the type of reactor and reactor controls to be used (Karlsen and Villadsen, 1984).

Conventional stirred-tank reactors are used with adaptations for immobilized biocatalyst retention or recuperation; a less shear-intensive agitation system is usually also needed to avoid mechanical damage to the enzyme or cell carrier as well as an efficient system for the addition and distribution of alkali to avoid local alkaline inactivation of enzyme and penicillin, and to obtain strict pH control (Figure 8). If this reactor type still fails because of biocatylyst-carrier limitations, a recycle reactor-column combination could be used. In such a system, pH adjustment occurs separately from the immobilized enzyme (Savidge, 1984).

Relatively little information is available on the acylation of 6-APA to produce (semisynthetic) penicillins using (immobilized) biocatalysts (Mc Dougall et al., 1982; Vandamme, 1983; Nam and Ryu, 1984). The penicillin G acylases from E. coli ATCC 9637, Bacillus circulans, B. megaterium and the ampicillin acylase from Pseudomonas melanogenum are able to synthesize ampicillim or amoxycillin from 6-APA and D-phenyl-glycine methyl ester, or p-hydroxyphenylglycine methyl ester, respectively.

Using ampicillin acylase, there is no need to separate the side chain, liberated during the 6-APA formation step. These acylases display a distinct preference, although not an absolute requirement, for the D-configuration of the side-chain molecule.

The development of a commercially feasible penicillin acylase process for enzymatic synthesis of semisynthetic penicillins from 6-APA is still unlikely due to the low conversion ratios, difficult product separation, high cost of the side-chain and the availability of relatively simple chemical acylation procedures.

B) Cephalosporin bioconversions

Cephalosporins are penicillin analogues that are synthesized by direct fermentation with Cephalosporium acremonium (Acremonium chrysogenum), or by chemical ring expansion of penicillin. The major fermentaion product, cephalosporin C, contains the 7-aminocepalosporanic acid (7-ACA) nucleus and the side chain A-aminoadipic acid (Figure 9). 7-ACA derivatives with other side chains have not been obtained by fermentation.

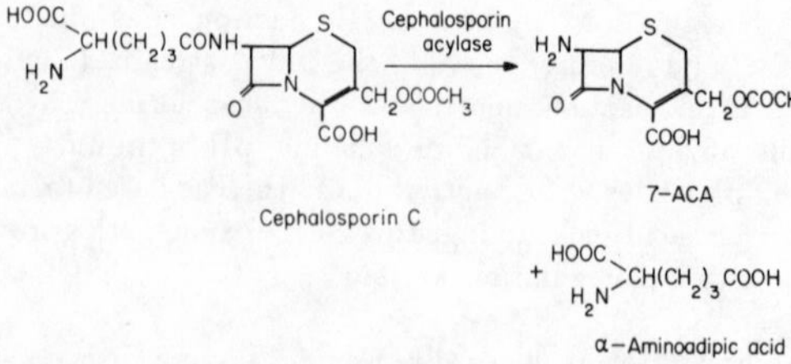

Figure 9 Deacylation of cephalosporin C. Conversion of cephalosporin C into 7-aminocephalosporanic acid (7-ACA) and α-aminoadipic acid by cephalosporin C acylase

Cloning of acyltransferase genes from the penicillin producer <u>Penicillium chrysogenum</u> into <u>Cephalosporium acremonium,</u> the cephalosporin C producer, might result in cephalosporin fermentaions yielding new precursored cephalosporins (see also Figure 4). The synthesis of semisynthetic cephalosporins containg the 7-ACA nucleus thus far depands upon chemical removal of the A-amnioadipic acid side chain.

Japanese patents claim the direct enzymatic hydrolysis of cephalosporin C into 7-ACA and α-aminoadipic acid with <u>Psuedomonas putida</u> cephalosporin C acylase. A two-step enzymic process for 7-ACA production has also been proposed by Fujil et al., (1979). Cephalsporin C is first transformed by <u>Trigonopsis variabilis</u> CBS 4095 or <u>Rhodotorula gracilis</u> D-amino acid oxidase (E.C. 1.4.3.3.) activity into keto-adipic-7-aminocephalosporanic acid, which spontaneously yeilds glutaryl 7-ACA, which is then hydrolysed by glutaryl-7-ACA acylase of <u>Comamonas or Pseudomonas</u> sp. into 7-ACA (Figure 10). This process has been commercialized recently by Asahi Chemicals Co., Tokyo, Japan. (Scheme 3 and 7).

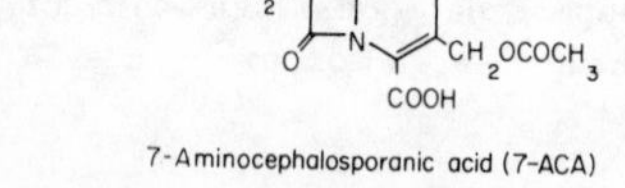

Figure 10 Bioconversion of 7-β-(4-carboxybutanamido)cephalosporanic acid (glutaryl 7-ACA) into 7-ACA by glutaryl 7-ACA-acylase

Toyo Jozo in Japan reported in detail on the isolation and properties of <u>Pseudomonas</u> (<u>Pseudomonas putida</u>, <u>Pseudomonas</u> SY-77-1) strains, able to deacylate 7-B-(4-carboxybutane-amido) cephalosporinic acid (glutaryl-7-ACA) into 7-ACA (Figure 10). These strains specifically hydrolysed cephalosporin-compounds having aliphatic dicarboxylic acid in the acyl side chain; cephalosporin C. was not hydrolysed (Shibuya et al., 1981; Ichikawa et al., 1981).

TWO-STEP BIO-PROCESS FOR 7 - ACA PRODUCTION

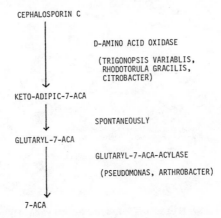

CEPHALOSPORIN C

D-AMINO ACID OXIDASE

(TRIGONOPSIS VARIABLIS,
RHODOTORULA GRACILIS,
CITROBACTER)

KETO-ADIPIC-7-ACA

SPONTANEOUSLY

GLUTARYL-7-ACA

GLUTARYL-7-ACA-ACYLASE

(PSEUDOMONAS, ARTHROBACTER)

7-ACA

Glutaryl-7-ACA acylase has also been detected in Bacillus, Arthrobacter, and Alcaligenes species (Inoue et al., 1979).

Direct fermentation of glutaryl-7-ACA rather than cephalosporin C formation and use of glutaryl-7-ACA-acylase could provide an alternative economic route for 7-ACA-production. Genetic transfer of the above mentioned Rhodotorula gracilis or Tvrigonopsis variabilis activity (Simonetta et al., 1982; Szwajcer and Mosbach, 1985; Kubicek-Pranz and Rohr, 1985) into cephalosporium would allow for such a direct fermentation of of glutaryl-7-ACA.

Immobilized enzyme or cell technology should soon be applied to this new type of acylase and it might compete with conventional chemical production menthods for 7-ACA, which can, in turn, be converted into the clinically important cephaloglycine and cephalothin (Figure 11). These coupling reactions have already been performed with the Celite-adsorbed penicillin G-acylase from B. megaterium and with Xanthomonas citri cells (Nam et al., 1983; Kim et al., 1983). An immobilized cell system has been described by Fukushima et al. (1976) where cells of Pseudomonas sp. or Comamonas sp. entrapped in cellulose acetate capsules hydrolysed glutaryl-7-ACA into 7-ACA. Another useful bioconversion is the deacylation of 7-ACA and cephalosporins (Figure 12). Acetate removal by enzymic hydrolysis (acetylesterase) is the only efficient method to prepare deacetyl 7-ACA.

Bacillus subtilis esterase absorbed onto bentonite was used for multiple batch reactions to deacetylate 7-ACA solutions (Abbott et al., 1976).

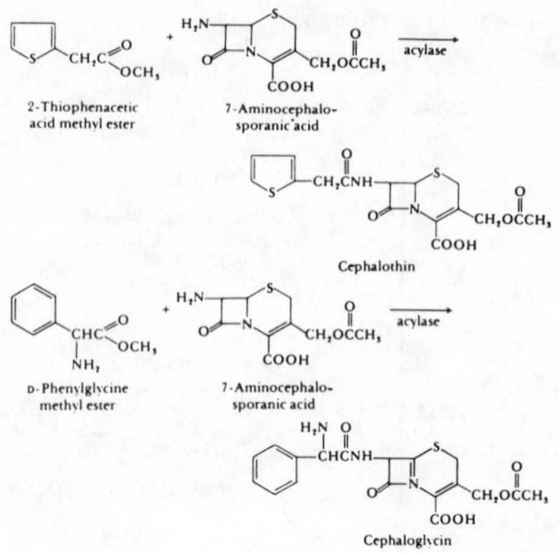

Fig. 11 Synthesis of Semisynthetic Cephalosporins from 7-ACA.

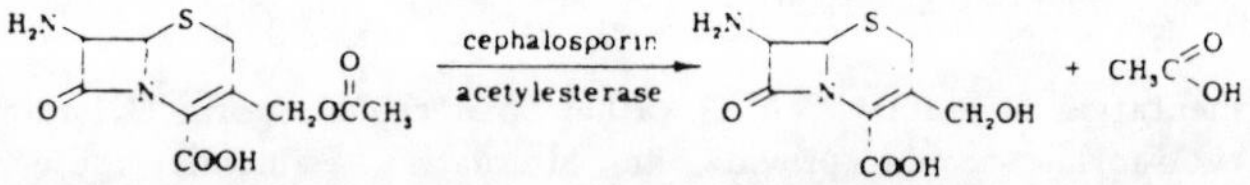

Fig. 12 Conversions of 7-ACA into deacetyl 7-ACA by Cephalosporin Acetylesterase.

Enzymes able to hydrolyse 7-ACA into its desacetoxy derivative 7-ACA (Figure 13) or to form 7-ACA through coupling of an acetoxy-group to 7-ACA have not yet been described.

Cephalosporins can also be produced from the precusor penicillin G or penicillin V by a series of chemical reactions that expand the five-membered thiazolidine ring of the penicillins into the six-membered dihydrothiazine ring of the cephalosporins. The cephalosporins thus obtained contain the penicillin G or V side-chain, which can then be removed enzymatically by conventional penicillin G or V-acylase action. The cephalosporin nucleus thus obtained is 7-aminodesacetocycephalosporanic acid 7-ADCA) (Figure 13). Direct conversion of 6-APA into 7-ADCA has not yet been reported.

Immobilized biocatalysts have been used for the deacylation (or acylation) of such cephalosporin compounds. Extracellular penicillin G acylases from E. coli, B. megaterium and P. rettgeri adsorbed onto Celite, activated carbon, carboxymethyl-cellulose or an Amberlite ion-exchange resin hydrolysed 7-phenyl-acetamide ADCA.

The partially purified penicillin-V acylase as well as intact cells of Erwinia aroideae were entrapped in cellulose triacetate fibres at Glaxo Laboratories, UK; a fibre-packed column hydrolysed 7-phenoxyacetamido-ADA to 7-ADCA in 58% yield after 3 h of substrate circulation at 37° C (Fleming et al., 1974). Acylation of 7-ADCA with D-phenylglycine derivatives produces the useful antibiotic cephalexin (Figure 14).

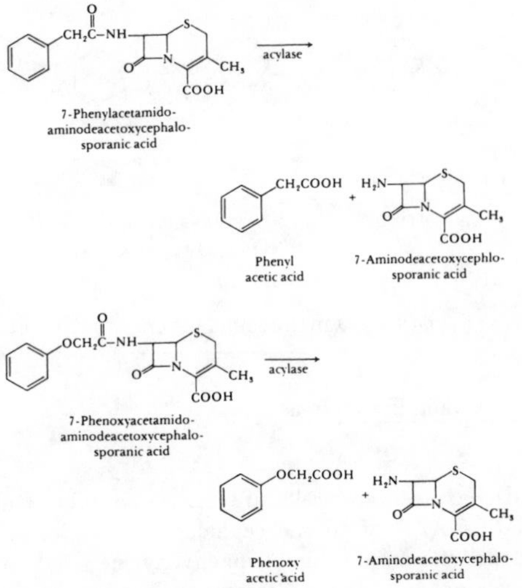

Fig. 13 Hydrolysis of Cephalosporins into 7-Aminodeacetoxy-Cephalosporic Acid (7-ADCA).

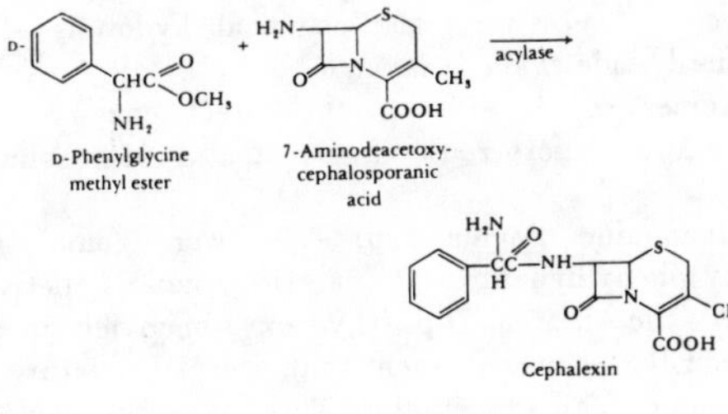

Fig. 14 Acylation of 7-ADCA into Cephalexin.

Several immobilized enzyme or cell systems for the synthesis of this clinically useful antibiotic have been developed (Kato et al., 1980; Takahashi et al., 1979; Kim et al., 1983; Vandamme, 1983). The enzyme from Acetobacter turbidans ATCC 9325 has been immobilized to curdlan and used for continuous synthesis of cephalexin (Takahashi et al., 1977). Whole cells of Xanthomonas citri, immobilized in carrageenan continuously produced cephlexin at pH 6.2 at 25° C with 83% conversion rate (Kim et al., 1983) (Figure 15).

In contrast to enzymic acylation, chemical acylation of the cephalosporin nucleus needs blocked derivatives of the reactants. This fact, combined with the advantages of immobilized enzyme or cell technology, favours the use of enzymic procedures in semi-synthetic cephalosporin synthesis.

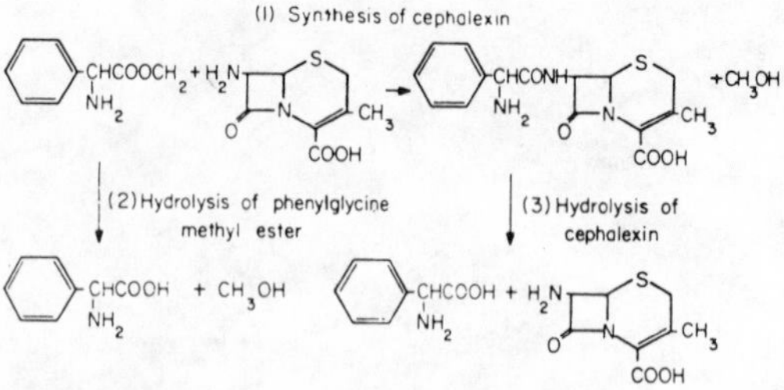

Fig. 15 Reactions Catalysed by Xanthomonas citri in the Case of Cephalexin.

C) *B*-lactam-antibiotic Side Chain Production

Microbial (immobilized) enzymes and cells are also used in the manufacture of side-chain acids to be coupled to 6-APA or to the cephalosporin nucleus. Enzymic resolution processes have been developed for producing D-phenylglycine used to prepare ampicillin, cephalexin and cephaloglycine, and D-p-hydroxyphenylglycine to prepare amoxycillin. These enzymatic processes have recently reached the industrial production stage.

Several enzymes, including E. coli penicillin G. acylase, Specific for the L-isomer-hydrolysis of N-phenylacetyl-DL-hydroxyphenylglycine or DL-phenylglycineamide, have been described; such processes necessitate the chemical hydolysis of the unaffected D-derivative into the desired side-chain (Savidge et al., 1974). It would thus be advantageous if the D-isomer could serve as the direct enzyme substrate, while the unaffected L-isomer is racemised before reuse (Sugie et al., 1980; Schutt, et al., 1985).

Pseudomonas and alkaophilic Bacillus sp. cells were found to stereospecifically hydrolyse DL-p- (hydroxy)phenylhydantoin to a D-carbamoyl derivative, which upon chemical hydrolysis yielded the desired D-p- (hydroxy)-phenylglycine; the remaining L-phenylhydantoin spontaneously racemized such that the DL-mixture was finally totally converted to the D-compound (Yamada et al., 1980). Further developments led to the isolation of strains(i.e. Agrobacterium radiobacter NRRL-B 11291) with both D-hydantoinase as well as D-carbamoylase activity, thereby avoiding the need for the complex chemical hydrolysis of the D-carbamoyl derivative (Olivieri et al., 1979). Such a process allows the complete conversion of the racemic substrate to the desired D-amino acid in a single reactor (Figure 16). A Flavobacterium hydantoinophilum strain is claimed to carry out these reactions and also to couple the side chain thus formed to 6-APA, with formation of amoxycillin in one single run! Enzymatic side chain resolution is carried out on an industrial scale by Kanegafushi G., Singapore, and by Recordati (Snam Progetti) in Italy. (Scheme 3).

D) Bioconversion of Other Antibodies.

The formation of the broad spectrum aminoglycoside anitibiotic amikacin from kanamycin A and L(-)*A*-hydroxy-*Γ*-aminobutyric acid with a mutant of the butirosin

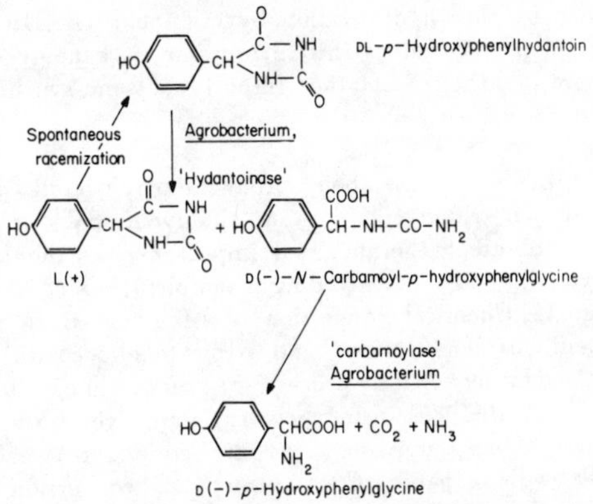

Fig. 16 Enzymatic Resolution of DL-ρ-hydroxyphenylhydantoin
into D-ρ-hydroxyphenylglycine with Hydantoinase and Carbamoylase
from Agrobacterium sp.

producing <u>Bacillus</u> <u>circulans</u> could replace its expensive chemical synthesis; furthermore a crude fermentation broth from <u>Streptomyces</u> <u>kanamyceticus</u> could be used as kanamycin A substrate for bioconversion (Cappelletti and Spagnoli, 1983) (Figure 17).

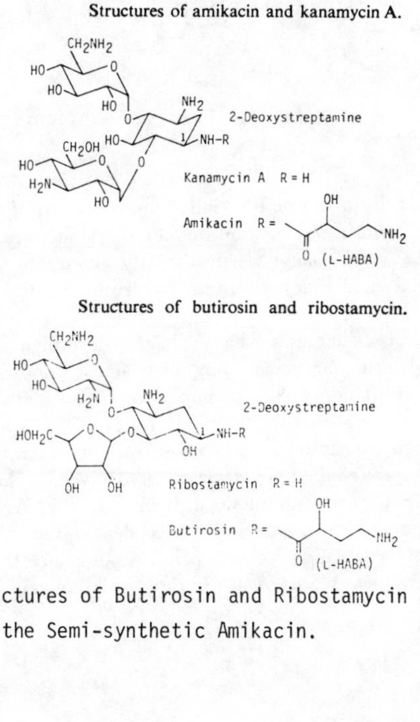

Structures of amikacin and kanamycin A.

Structures of butirosin and ribostamycin.

Fig.17 Structures of Butirosin and Ribostamycin and of Kanamycin A
and the Semi-synthetic Amikacin.

Alternatively, cloning of this hydroxyaminobytyric acid (HABA)-acylase gene from the butirosin-producing Bacillus circulans into Streptomeces kanamyceticus could lead to the direct and efficient synthesis of the important semi-synthetic aminoglycoside, amikacin.

Rifamycin B, one of the major metabolic products of Nocardia mediterranei, can be converted into rifamycin O, rifamycin S and rifamycin SV by chemical methods. Especially, rifamycin S is of great therapeutical importance as the key intermediate for the synthesis of rifamycin-derivatives, including rifampicin, used as an antituberculosic and antilepra-drug (Fig. 18).Chemical conversion of rifamycin B to rifamycin S requires an organic solvent system, artificial oxidant and strong acidic conditions, which result in unsatisfactory yields. A bioconversion process of rifamycin B to rifamycin S was recently described by Lee et al.(1985), using polyacrylamide gel entrapped acetone-treated cells of Humicola sp. and Monocillium sp. in a fluidised bed reactor at pH 7.8 and 50° C. The Humicola cells were capable of catalysing the conversion of rifamycin B to rifamycin O by rifamycin B-oxidse action; the thus formed rifamycin O was spontaneously hydrolysed to rifamycin S in neutral aqueous conditions. It was found that aeration, cell leakage and precipitation of rifamycin S were important factors in the continuous reaction.

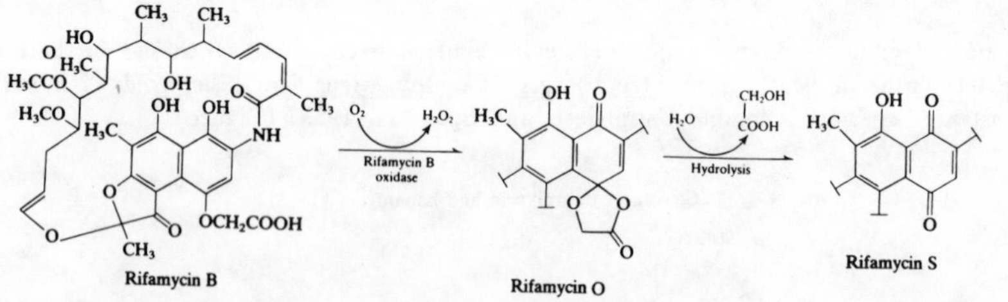

Fig. 18 Proposed Mechanism of Rifamysin Oxidase Reaction.

The macrolide antibiotic tylosin , commercially produced by Streptomyces fradiae, can be converted into 4-acyltylosin by Streptomyces thermotolerans; cloning of this enzyme into S. fradiae would yield 4-acyltylosin directly; co-immobilization of the two Streptomyces species could also yield 4-acytylosin in one run. (Fig.6).

A range of new natural B-lactams have been described recently, including cephamycins, clavulanic acid, thienamycins, olivanic acids, nocardicins, formadicins, cephabaccins, carpetimycins, asparenomycins, monobactams and chitinovorins (Imada et al., 1981; Sykes et al., 1981; Parker et al., 1982; Kitano, 1983). Many of these natural compounds have been reported to undergo bioconversion by microbial cell or enzyme preparations. Pseudomonas schuylkillensis acylase hydrolysed the single B-lactam nocardicin C at pH 8 and 37° C into 3-aminonocardicinic acid (3-ANA) (Komori et al., 1978). The olivanic acid related PS-5 antibiotic was deacylated by Pseudomonas sp. 1158 cells, immobilized in polyacrylamide gel at pH 7,4 and 30° C (Fukugawa et al., 1980). An E. coli acylase has been reported to acylate thienamycin reversibly (Okachi, 1979).

E) Bioconversion of Monobactams.

Monobactams are monocyclic N-sulfonated *B*-lactams isolated from bacteria (Pseudomonas, Gluconobacter, Chromobacter, Agrobacterium) (Sykes et al., 1981): they are N-acyl derivatives of the nucleus 3-aminomonobactamic acid (3-AMA) (Fig.19). Natural monobactums did not serve as substrates for E. coli penicillin acylase; the chemically prepared nucleus 3-AMA as well as 3-amino-4α-methylomonobactamic acid could be acylated at pH 4,5 at 45° C using a polyacrylamide immobilized E. coli penicillin G acylase. Unnatural acyl residues which could be coupled were readily deacylated at pH 7,5 (O'Sullivan and Aklonis, 1984). A further screening for specific monobactam acylase could lead the way to immobilized biocatalyst technology for semi-synthetic monobactam-production.

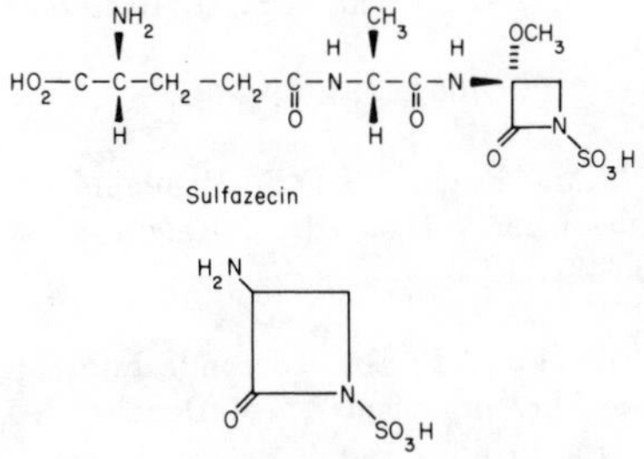

Fig. 19 Structures of the Monobactam Sulfazecin and of its Nucleus, 3-Aminomonobactamic Acid (3-AMA).

CONCLUSION

Contrary to immobilzed mono-step bioconversions for antibiotic production, which are already operational on a large scale (6-APA and 7-ACA production, side chain resolution), industrial biocatalysis by multi-step enzyme systems or by immobilized cell fiagments or viable cells remains a goal to achieve. The complexity of the pathways involved and the many basic aspects yet to be well understood hamper a quick translation of this concept into economic reality. Indeed, microbial, physiological and mechnical problems inherent to the immobilized biocatalyst principle, need further study to arrive at an optimal performance of immobilized catalyst bioreactors.

For most complex multi-enzyme processes, there is as yet no economic incentive to replace well-established fermentation processes by immoblilized enzyme or cell reactor technology; for bioconversion reactors, emphasis is clearly focused upon switching from free or immobilized single-enzyme reactions to immobilized (viable) cell processes.

Cloning of genes, coding for antibiotic-bioconversion enzymes such as: cyclase, penicillin G and V acylase (Schomer et al., 1984; Olsson et al., 1985), ampicillim-acylase, cephalosporin C-acylase, glutaryl-7-ACA-acylase (Matsuda and Komatsu, 1985) hydantoinase, ... should receive more attention so as to produce them more efficiently and abundantly: otherwise such genes could be introduced into antibiotic producer strains so as to extend or prolong their antibiotic pathways with an additional useful last step reation (acyl-transferase, HABA aclyase, tylosin-acylase,...) (Vournakis and Elander, 1983) and to obtain at the end of the fermentation process the semi-synthetic derivative directly!

LITERATURE

1. Abbott, B.J., Cerimele, B. and Fukuda, D.S.; Biotechnol. Bioeng., 19: 1033, 1976.

2. Cappelletti, L.M. and Spagnoli, R.; J. Antibiotics, 36: 328, 1983.

3. Chibata, I. and Tosa, T.; Annu. Rev. Biophys. Bioeng., 10: 197, 1981.

4. Daumy, G.O., Danley, D. and McColl, A.S.; J. Bacteriol, 163: 1279-1281, 1985.

5. Demain, A.L.; Science, 219: 709, 1983.

6. Demain, A.L., Piret, J.M., Friebel, T.O.E, Vandamme, E.J. and Matteo, C.C.; In "Microbiology" (Schlessinger, D., ed.), American Society for Microbiology, Washingtion, D.C.: p. 437, 1976.

7. Demain, A.L. and Wang, D.I.; In "Second International Symposium on the Genetics of Industrial Microorganisms", (McDonald, K.D., ed.), Academic Press, New York-London; p.115, 1976.

8. Deo, Y.M. and Gaucher, G.M., Biotechn Letters, 5(2): 125, 1983.

9. Egorov, N.S., Baranova, I.P. and Kozlivoa, Y.U.I.; Antibiotiki (Moscow); 23: 872-874, 1978.

10. Fleming, I.D., Turner, M.K. and Napier, E.J.; Ger. Pat.; 2 422 374, 1974.

11. Foster, B.C., Coutis, R.T., Pasutto, E.M., and Dossetor, J. B., Biotechnol. Letters; 5(10): 693, 1983.

12. Freeman, A. and Aharonowitz, Y.; Biotechnol. Bioeng.; 23: 2747, 1981.

13. Fujii, T., Shibuya, T. and Matsumoto, K.; Proc. Annu. Meet. Agric. Chem. Soc., Japan; 1-4: April, 1979.

14. Fukugawa, Y., Kuno, K., Ishikura, T. and Kouno, K.; J. Antibiotics; 33: 543, 1980.

15. Fukui, S. and Tananka, A, Annu. Rev. Microbiol.; 36: 145, 1982.

16. Fukushima, M., Fujii, T., Matsumoto, K. and Morishita, M.; Japan Pat.; 7: 670 884., 1976.

17. Gasson, M. J.; FEMS Microbiol. Letter; 21: 7, 1984.

18. Green, C. J. Diagnostic Histopath; 4: 157, 1981.

19. Harrison, F.G. and Gibson, E.D.; Proc. Biochem.; Feb., 33, 1984.

20. Hurst, A.; Adv. Appl. Microbiol.; 27: 85, 1981.

21. Ichikawa, S., Shibuya, Y., Matsumoto, K., Fujii, T., Komatsu, K. and Kodaira, R.; Agric. Biol. Chem.; 45: 2231-2236, 1981.

22. Imada, A., Kitano, K., Kintaka, K., Muroi, M., and Asai, M.; Nature; 289: 590-591, 1981.

23. Inoue, T., Matsuda, K., Fukuo, T. and Kawate, S.; Abstr. Annu. Meet. Agric. Chem. Soc.; Japan Tokyo; p. 220, April 1979.

24. Jensen, S.E., Westlake, D.W.S., Bowers, R.J. and Wolfe, S.; J. Antibiotics.; 35: 1351, 1982.

25. Jensen, S.E., Westlake, D.W.S. and Wolfe, S.; Appl. Microbiol. Biotechnol.; 20: 155, 1984.

26. Karlsen, L.G. and Villadsen, J.; Biotechnol. Bioeng.; 26: 1485, 1984.

27. Karube, I., Suzuki, S. and Vandamme, E.J.; In "Biotechnology of Industrial Antibiotics Inc.", (E.J.Vandamme, Marcel Dekker, eds.), USA; p.761, 1984.

28. Kasche, V., Haufler, U. and Zolliner, R.; Hoppe Seyer's Z. Physiol.Chem.; 365: 1435, 1984.

29. Kato, K., Kawahata, K., Takahashi, T. and Kakinuma, A.; Agric. Biol. Chem.; 44: 1069, 1980.

30. Kim, J.K., Nam, D.H. and Ryu, D.D.Y.; Appl. Biochem. Biotechnol.; 8: 195, 1983.

31. Kitano, K.; Prog. Indust. Microbiol.; 17: 37, 1983.

32. Kleinkauf, H. and Koischwitz, H.; In "Multifunctional Proteins", (H. Bisswanger and E. Schminke-Ott., eds.), J. Wiley and Sons, Inc., New York, p. 217, 1980.

33. Klibanov, A.M.; Science; 219: 732, 1983.

34. Komori, T., Kunugiya, K., Nakahara, K., Aoki, H. and Imanaka, H.; Agr. Biol. Chem.; 42: 1439, 1978.

35. Krause, M., Marahiel, M.A., Von Dohren, H., Kleinkauf, H.; Bacteriol; 162: 1120-1125, 1985.

284

36.   Kubicek-Pranz, E.M. and Rohr, M.; Biotechnol. Letters; 7: 9, 1985.

37.   Kurzatkowski, W., Kurylowics, W. and Paszkiewics, A.; Eur. J. Appl. Microb. Biotechnol.; 5: 211, 1982.

38.   Lee, G.M., Choi, C.Y., Park, J.M. and Han, M.M.; J. Chem. Technol. Biotechnol.; 358: 3, 1985.

39.   Lowe, D. S.; Dev. Ind. Microbiol.; 26: 143, 1985.

40.   Luengo, J.M., Alemany, M.T., Salto, F., Ramos, F., Lopez-Nietro, M. J. and Martin, J.; Bio/Technology; 4: 44-47, 1986.

41.   Madry, N., Zocher, R., Grodzki, K. and Kleinkauf, H.; Appl. Microbiol. Biotechnol.; 20: 83, 1984.

42.   Mahajan, P. B.; Appl. Biochem. Biotechnol.; 9: 537, 1984.

43.   Mahajan, P. B. and Borkar. Appl. Biochem. Biotechnol.; 9: 421, 1984.

44.   Marconi, W., Cecere, F., Morisi, F., Della Penna, G. and Rappuoli, B.; J. Antibot.; 26: 228, 1973.

45.   Matsuda, A. and Komatsu, K. J. Bacteriol.; 163: 1222-1228, 1985.

46.   McDougall, B., Dunnill, P. and Lilly, M.D.; Enz. Microb. Technol.; 4: 114, 1982.

47.   Morikawa, Y., Karube, I. and Suzuku, S.; Biotechnol. Bioeng.; 22: 1015, 1980.

48.   Morikawa, Y., Karube, I. and Suzuki, S.; Biotechnol. Bioeng. 21: 261, 1979a.

49.   Morikawa, Y., Ochiai, K., Karube, I. and Suzuki, S.; Antimicrob. Agents Chemother.; 15: 126, 1979b.

50.   Nam, D. H. and Ryu, D. D. Y. J. Antibiotics; 37: 1217, 1984.

51.   Nam, D.H., Sohn, H.S. and Ryu, D.D.Y.; Bull. Korean Chem. Soc.; 4: 72, 1983.

52.   Okachi, R. J.; Agric. Chem. Soc. Japan. 53: R169, 1979.

53.   Olivieri, R., Fascetti, E., Angelini, L. and Degen, L.; Enzyme Microb. Technol.; 1: 201, 1979.

54.   Olsson, A., Hagstrom, T., Nilsson, B., Ulem, M. and Gatenbeck, S. Appl. Environm. Microbiol. 49: 1084, 1985.

55.   O'Sullivan, J. and Aklonis, C.A.; J. Antibiotics; 37: 804, 1984.

56. Parker, W.L., Rathnum, M.L., Wells, J.S., Treso, W.H., Principle, P. A. and Sykes, R. B.; J. Antiobiotics.; 35: 653, 1982.

57. Poirier, A. and Demain, A. L. Antimicrob. Ag. Chemother.; 20: 508, 1981.

58. Poulsen, P. B. Biotechn. Gen. Eng. Revs.; 1: 121, 1984.

59. Rousseau, I., Roth-Van Eijk, C. and Liou, J.K.; Progress Industrial Microbiol.; 20: 73, 1984.

60. Ryu, D. D. Y. and Nam, D. H. Enz. Eng. News. 9: 30, 1983.

61. Savidge, T.A.; In "Biotechnology of Industrial Antibiotics" (E. J. Vandamme, ed.), Marcel Dekker Inc.; p. 171, 1984.

62. Savidge, T.A., Powell, L.W. and Lilly, M.D.; Br. Pat.; 1: 357, 317, 1974.

63. Schomer, U., Segner, A. and Wagner, F.; Appl. Environm. Microbiol.; 47: 307, 1984.

64. Sebek, O.K.; In "Economic Microbiology", (A. H. Rose, ed.), Academic Press, New York, London; Vol.5: 575, 1980.

65. Shibuya, Y., Matsumoto, K. and Fujii, T.; Agric. Biol. Chem.; 45: 1561-1567, 1981.

66. Schutt, M., Schmidt-Kastner, G., Arens, A. and Preiss, M.; Biotechnol. Bioeng.; 27: 420, 1985.

67. Simonetta, M.P., Vanoni, M.A. and Curti, B.; FEMS Microbiol. Lett.; 15: 27, 1982.

68. Sugie, M. and Suzuki, H. Agric. Biol. Chem.; 44: 1089, 1980.

69. Sykes, R.B., Bonner, D.P., Bush, K., Georgopapadaku, N.H. and Well, S.S.; J.Antimicrob. Chemother.; 8: (Suppl.E.), 1-16, 1981.

70. Szwajcer, E. and Mosbach, K. Biotechnol. Letters; 7: 1, 1985.

71. Takashashi, T., Kato, K., Yamazaki, Y., Isono, M.; J. Antiobiot.; 30: 130, 1977.

72. Vandamme, E. J. Chem. Ind.; 24: 1070, 1976.

73. Vandamme, E. J. In "Economic Microbiology", (A. H. Rose, ed.), Academic Press, New York, London; 5: 467, 1980.

74. Vandamme, E. J. In "Tropics in Enzyme and Fermentation Biotechnology", (A. Wiseman, ed.), Ellis Horwood, Chichester; 5: 185, 1981.

75.   Vandamme, E. J. Enz. Microb. Technol.; 5: 403, 1983.

76.   Vandamme, E.J. (ed.); Biotechnology of Industrial Antibiotics; Marcel Dekker Inc., New York; p. 832, 1984.

77.   Veelken, M. and Pape, H. Eur. J. Appl. Microb. Biotechn.; 15: 206, 1982.

78.   Venkatasubramanian, K. and Vieth, W.R.; In "Prog. Ind. Microbiol."; 15: 61, 1979.

79.   Vournakis, J. N. and Elander, R. P.; Science; 219: 703, 1983.

80.   Wang, D.I.C., Meir, J. and Yokoyama, K.; Appl. Biochem. Biotechnol.; 9: 105-116, 1984.

81.   Wolfe, S., Demain, A.L., Jensen, S.E. and Westlake, D. W. S. Science; 226: 1386, 1984.

82.   Yamada, H., Shimizu, S., Shimada, H., Tani, T., Takashashi, S. and Ohashi, T. Biochimie; 62: 395, 1980.

PRODUCTION OF ORGANIC ACIDS BY IMMOBILIZED CELLS OF FUNGI

H. Horitsu, Y. Takahashi, S. Adachi, R. Xioa, T. Hayashi and K. Kawai

Applied Microbiology
Department of Agricultural Chemistry
Faculty of Agriculture
Gifu University
Gifu 501-11, Japan

H. Kautola
Biotechnology and Food Engineering
Department of Chemistry
Helsinki University of Technology
Espoo, Finland

ABSTRACT

Production of citric-, itaconic-, and L-lactic-acids using one or multiple stages, or series-types of bioreactors by immobilized cells of Asp. niger. Asp. terrus and Rh. oryzae which are respectively entrapped into polyacrylamide gel or Ca-alginate gel as carriers are researched.

Among some chemicals tested, magnesium sulfate, ammonium nitrate and calcium carbonate had a positive effect on production of citric, itaconic and L-lactic acids, respectively. As for the other factors enhancing the acid productivity, oxygen tension and shape of immobilized cells are also important.

Repeated and continuous uses of the immobilized cells were successfully achieved when the fungal cells could moderately proliferate within gels under optimized conditions.

Acid production rates of 25 mg of citric acid/hr/10 g gel having a half life time of 105 days, 28 mg of itaconic acid/hr/10 g gel having a half life time of 30 days, and 52 mg of L-lactic acid/hr/ 10 g gel having a half life time of 17 days, were observed in the respective continuous methods.

INTRODUCTION

Citric acid and lactic acid are used as food additives, and itaconic acid is used as a starting material for plastics.

Recently, there have been many reports on the application of immobilized cell technology for the biotechnical production of organic compounds; that is, organic acids, antibiotics, enzymes, steroids and ethanol have been produced by immobilized microbial cells, but, bacteria and yeasts are mainly used while the use of fungi is limited.

Fungi are used in wide fields of the fermentation industry; for example, production of penicillin, gibberellin, citric acid, itaconic acid, lactic acid and other organic compounds is carried out by fungi. Moreover, many enzymes including, α-amylase, protease and lipase are also produced by fungi.

There is considerable interest in improving the productivity and product economy of organic products by using immobilized cells of fungi. Among such organic products, production of organic acids, citric acid, itaconic acid and lactic acid with immobilized cells of fungi are very economically interesting.

Before we started our research on citric acid production using immoilized cells of Asp. niger, there was only one report on citric acid production with immobilized cells of Asp. niger; that is, Vieth and Venkatasubramanian (14) reported that cells of Asp. niger were dispersed in 4% hide collagen solution to form a membrane which was crosslinked with glutaraldehyde. Then the immobilized cells were used to produce citric acid from a sucrose solution.

But, there are some problems in this report because this type of immobilization is difficult to prepare due to membrane form and citric acid productivity being lower than conventional fermentation, and the activity being only about 6-7 days.

We have been researching from twenty years ago on citric acid production using a tower fermentor and recently changed the production of citric acid with more saving method for material and energy by using immobilized cells of Asp. niger. We also have applied this method to other organic acids, itaconic acid, lactic acid and qluconic acid.

In this report, we describe the results of citric acid, itaconic acid and lactic acid production by other researchers and us.

EXPERIMENTAL

Microorganisms

Asp. niger G-011, Asp. terrus G-026 and Rh. oryzae AHU 6536 were used for citric acid, itaconic acid and lactic acid production respectively.

<u>Culture media</u>

For citric acid, beet molasses (sucrose, 10%), pH adjusted to 5.8, was used, and for itaconic acid, a mixture of 10% glucose, 0.37% NH_4NO_3, 0.2% $MgSO_4$-$7H_2O$, 0.27% and $NH_4H_2PO_4$ and 0.1% corn steep liquor, pH 5.0 and for lactic acid, a mixture of 5% glucose, 0.05% NH_4NO_3, 0.03% KH_2PO_4, .025% $MgSO_4$-$7H_2O$ and 2% $CaCO_3$, pH7.15, was used.

<u>Cultivation.</u>

Each spore suspension (5×10^7 of <u>Asp. niger</u>, <u>Asp. terreus</u> or <u>Rh. oryzae</u>) was inoculated into 100-ml of each medium in 500-ml shaking flasks and cultivated on a shaker (7 cm stroke, 120 reciprocations /min) at 30-35°C. Grown cells were harvested at indicated times.

BIOREACTOR

Two types of bioreactors were used: 1. Series type reactor, as shown in Fig. 1, was used by one or multiple vessels connected in series. 2. Stage type reactor, as shown in Fig. 2, was used by one or multiple vessles connected in a pile.

Fig. 1. Series type of bioreactor   Fig. 2. Stage type of bioreactor

<u>Bioreactor substrates.</u>

For citric acid, the bioreactor substrate was composed of 10% sucrose, 0.02% NH_4NO_3, 0.5mg% $MgSO_4$.$7H_2O$,pH 3.0. For itaconic acid, the bioreactor substrate was composed of 6% glucose, 0.3% NH_4NO_3 and 0.2% $MgSO_4$.$7H_2O$,pH2.25. For lactic acid, the bioreactor substrate was composed of 5% glucose, 0.05% NH_4NO_3,0.025% $MgSO_4$ $7H_2O$ and 2% $CaCO_3$, pH 7.15.

290

Analytical method.

Citric acid, itaconic acid and lactic acid concentrations were determined by Shimaduz Isotacophoresis apparatus IP-2A using 0.01Mβ-alanine, 0.2% Triton X-100, pH 3.1 as the leading electrolyte and 0.01M n-caproic acid as the terminal electrolyte.

Immobilization of living cells.

The methods of immobilization of cells are shown in Figs. 3,4,5 and 6 respectively.

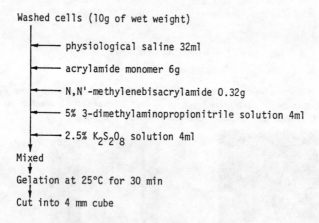

Fig. 3. Immobilization of cells with polyacrylamide

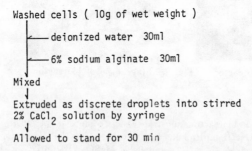

Fig. 4. Immobilization of cells with calcium alginate

291

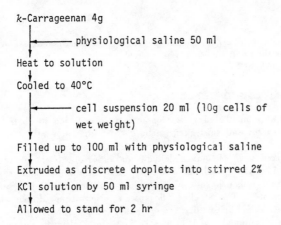

Fig. 5. Immobilization of cells with κ-carrageenan

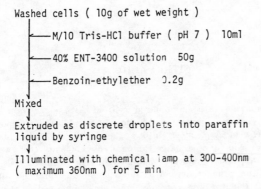

Fig. 6. Immobilization of cells with photocrosslinkable resin.

Acid production with immobilized cells.

Repeated batch method.

About 20g of gel cubes or beads were suspended in a glass reactor of 400cm^3 (5cm in diameter, 20cm in height) containing 100-ml of each bioreactor substrate.

Incubation was carried out under aeration at a rate of 0.7 1/min to 2.8 1/min at 30-35°C. The culture medium was replaced at 24-hr intervals.

Continuous method.

About 40g of immobilized cells were incubated in the same reactor as was used for the repeated batch method. Incubation was carried out at 30°C. with aeration at 1.4 l/min at a flow rate of 4 ml/hr.

Preparation of samples for electron microscope.

Immobilized cells were incubated in a bioreactor. After being incubated, the cubes were taken from the reactor and subsequently sliced to ca.1mm in width.

The slices were fixed in 7.5% glutaraldehyde containing 0.1M phosphate buffer, pH 7.3 at 0°C. for 1 hr. After washing with the phosphate buffer, the samples were dehydrated with ethanol (50,80,90,95% and absolute). The dehydrated samples were stocked in n-buthylacetate. The stocked slices were dried using Hitachi HCP-critical point drying apparatus. The dried slices were coated with carbon, followed with gold. The coated slices were examined by JEOL-JSM-U$_3$-scanning electron microscope of the Japan Electron Optics Laboratory.

RESULTS AND DISCUSSION.

We describe only the results having significant influence on the respective acid productions.

CITRIC ACID.

For citric acid production with immobilized cells of Asp. niger, in addition to the report by Vieth and Venkatasabramanian [14] Heinrich and Rehm[2] reported that by adsorption of the mycelia of Asp. niger onto a glass-carrier in a fixed-bed reactor, a small amount of citric acid and a large amount of gluconic acid were produced. Moreover, Eikmeier, et al.,[1], also reported that by using immobilized cells entrapped in Ca-alginate gel beads, the productivity of citric acid in the presence of 10-20 mg% NH_4NO_3 in an air-lift fermentor was twice as much that obtained in ordinary shaking culture. In addition Vaija, et al.,[13], reported that by using immobilized cells entrapped in Ca-alginate gel beads in an air-lift stirred reactor, 1.2% citric acid was obtained. Recently, we also obtained a good yield of citric acid production using immobilized cells entrapped in Ca-alginate gel beads [4].

On the other hand, we also reported that citric acid production was carried out using immobilized cells of Asp. niger by entrapment in polyacrylamide gel [3 5]. We would like to describe results obtained by using immobilized cells entrapped in polyacrylamide gel. Repeated batch method.

Effect of immobilization method on acid productivity.

A comparison of the immobilization method with acid productivity was carried out. Both the acrylamide method and alginate method gave good results for citric acid production. (Fig. 7).

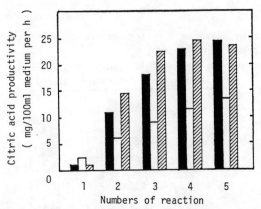

■; polyacrylamide, □; κ-carrageenan,▨; Ca-alginate.

Fig. 7. Effect of immobilization method on citric acid productivity

Effect of shape of immobilized cells.

The effect of the shape of immobilized cells on citric acid productivity was studied using either a cube or slice. Slice type of gels are better than cubic type of gels on yield of citric acid. It may be dependent on surface area and permeability of gels. (Fig. 8).

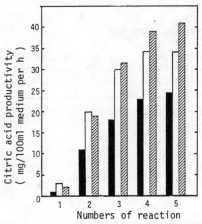

■; 4cubic mm, □; 2.5cubic mm,▨; 2.5mm slice (2.5x2.5x1.0mm).

Fig. 8. Effect of shape of immobilized cells on productivity

Effect of aeration volume.

The effect of aeration volume on citric acid production was investigated by changing the aeration volume from 0.7 l/min to 2.8 l/min.

The more the aeration was increased, the more citric acid was produced up to 1.4 l/min, but at the higher level, the effect was not observed. (Fig. 9).

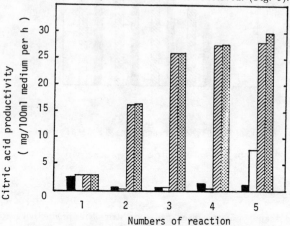

■; 0.7 l/min,□; 1.0 l/min, ▨; 1.4 l/min,▧; 2.8 l/min.

Fig. 9.　Effect of aeration volume on citric acid productivity

Continuous method.

Citric acid production using a two-stage bioreactor.

Citric acid production was investigated using a two-stage bioreactor. The maximum rate of citric acid production, 96.6 mg/100-ml medium/hr was observed after 8 days, and the period of half-life was found to be about 105 days. (Fig. 10).

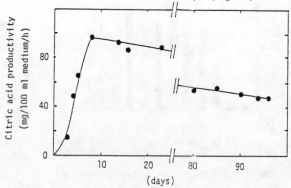

Fig. 10.　Continuous production of citric acid using two-stage bioreactor.

Electron microscope observation of immobilized cells.

To observe the state of cells in or on the gels with incubation, slices of gel cube were prepared at zero degrees. After 5 days of incubation they were submitted to scanning electron microscopy. Changes in the population of cells on the surface were observed with increasing incubation time. (Fig. 11).

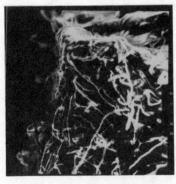 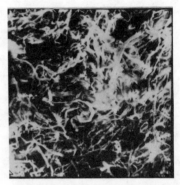

Initial cells(x600)   5 days-incubation cells(x600)

Fig. 11. Scanning electron micrographs on the surface of immobilized cells

ITACONIC ACID

For itaconic acid production with immobilized cells of Asp. terreus, Kautola et al.,[12] reported that by using immobilized cells of Asp. terreus entraped on cellite R-626 or in agar gel cubes, production of itaconic acid from glucose or xylose was carried ou, and productivities of 1.2 g/hr of the acid from glucose and 0.56 g/hr from xylose were obtained. Recently, Ju and Wang [11], also reported that immobilized cell disk bioreactor had advantages over other immobilized cell bioreactors. But we[6] [7] would like to describe results obtained using immobilized cells entrapped in polyacrylamide gel.

Repeated batch method

Effect of nitrate.

The effect of nitrate on itaconic acid productivity was investigated at concentrations of 0%, 0.15%, 0.3% and 0.6% nitrate. The highest rate of itaconic acid production was obtained by the addition of 0.3 - 0.6%. (Fig. 12).

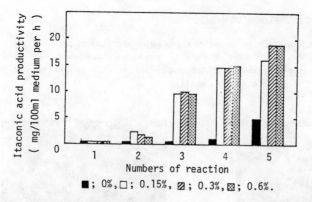

Fig. 12.   Effect of NH_4NO_3 concentration on itaconic acid productivity.

Effect of magnesium ion.

The effect of magnesium ion on itaconic acid productivity was investigated at concentrations of 0.025%, 0.05%, 0.1% and 0.2% $MgSO_4 \cdot 7H_2O$. The maximum rate of itaconic acid production was obtained by the addition of 0.05%. (Fig. 13).

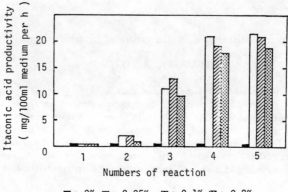

Fig. 13.   Effect of $MgSO_4$ $7H_2O$ concentration on itaconic acid productivity

Continuous method.

Itaconic acid production was investigated using a one-stage type of bioreactor. The maximum rate of itaconic acid production, 80 mg/hr, was observed after 15 days, and the period of half-life was found to be 30 days. (Fig. 14).

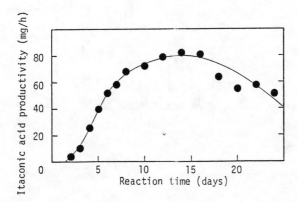

Fig. 14. Continuous production of itaconic acid using one-stage bioreactor.

LACTIC ACID

For lactic acid production with immobilized cells of fungi, there have been no reports. We[8] [9] [10] reported L-lactic acid production using immobilized cells of Rhizopus oryzae by entrapment in Ca-alginate.

Repeated batch method.

Effect of nitrate.

The effect of nitrate on lactic acid productivity was investigated at concentrations of 0%, 0.05% and 0.075% nitrate. The highest rate of lactic acid production was obtained by the addition of 0.05%. (Fig. 15).

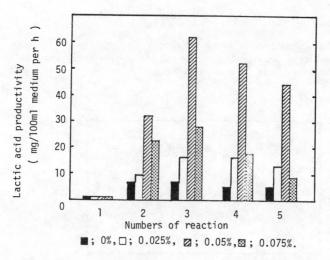

Fig. 15. Effect of NH_4NO_3 concentration on lactic acid productivity.

Effect of magnesium ion.

The effect of magnesium ion on lactic acid productivity was investigated at concentrations of 0%, 0.01%, 0.025% and 0.04% $MgSO_4$ $7H_2O$. The maximum rate of lactic acid production was obtained by the addition of 0.025%. (Fig. 16).

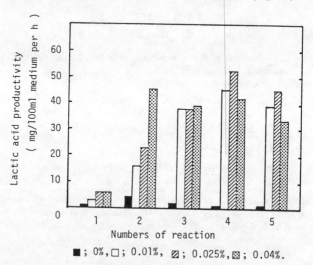

Fig. 16. Effect of $MgSO_4$ $7H_2O$ concentration on lactic productivity.

Continuous method

Lactic acid production was investigated using one-series type of bioreactor under controlled pH. The maximum rate of L-lactic acid production, 156 mg/hr, was observed after 2 days, and the period of half-life was found to be 17 days. (Fig. 17).

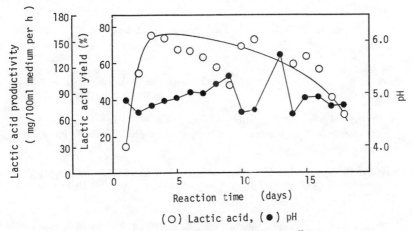

(O) Lactic acid, (●) pH

Fig. 17. Continuous production of lactic acid using pH controller

REFERENCES

1. H. Eikmeier, H., Westmeier, F. and Rehm, H.J.; Morphological development of Aspergillus niger immobilized in Ca-alginate and k-carrageenan, Appl. Microbiol. Biotechnol.; 19, 53-57, 1984.

2. Heinrich, M. and Rehm, H.J.; Formation of gluconic acid at low pH-values by free and immobilized Aspergillus cells during citric acid fermentation, Eur. J. Appl. Microbiol. Biotechnol.; 15, 88-92, 1982.

3. Horitsu, H.; Citric acid production with immobilized cells of Aspergillus niger; Japan Kokai; 1981/101345.

4. H. Horitsu, Takahashi, Y., Nakayama, Y., Kawai, K. and Kawano, Y.; Citric acid production with immobilized cells of Aspergillus niger by entrapment in Ca-alginate,;Abst. Ann. Meet. The Agr. Chem. Soc. Japan; p. 521, 1984.

5. Horitsu, H., Adachi, S., Takahashi, Y., Kawai, K. and Kawano, Y.; Production of citric acid by Asp. niger immobilized in polyacrylamide gels. Appl. Microbiol. Biotechnol. 22, 8-12, 1985.

6. Horitsu, H. and Kawai, K.; Itaconic acid production with immobilized cells of Asp. terreus, Japan Patent; 1982/173163.

300

7. Horitsu, H., Takahashi, Y., Tsuda, J., Kawai, K. and Kawano, Y.; Production of itaconic acid by Asp. terreus immobilized in polyacrylamide gels, Eur. J. Appl. Microbiol. Biotechnol., 18, 358-360, 1983.

8. Horitsu, H. and Kawai, K.; Lactic acid production with immobilized cells of Rh. oryzae, Japan Kokai.; 1983/111943.

9. Horitsu, H., Hayashi, T. and Kawai, K.; Production of L-lactic acid by Rh. oryzae immobilized in Ca-alginate, Abst. Ann. Meet. The Agr. Chem. Soc.Japan.; p. 664, 1985.

10. Horitsu, H., Hayashi, T. and Kawai, K.; Production of L-lactic acid by Rh. oryzae immobilized in polyacrylamide gels.; submitted to Appl. Microbiol. Biotechnol.

11. Ju, N. and Wang, S.S.; Continuous production of itaconic acid by Aspergillus terreus immobilized in a porous disk bioreactor,; Appl. Microbiol. Biotechnol.; 23, 311-314, 1986.

12. Kautola, H., Vahvaselka, M., Linko, Y.Y. and Linko, P.; Itaconic acid production by immobilized Aspergillus terreus from xylose and glucose.; Biotechnol.Lett.; 7, 167-172, 1985.

13. Vaija, J., Linko, Y.Y. and Linko, P.; Citric acid production with alginate bead entrapped Asp. niger ATCC 9142.; Appl. Biochem. Biotechnol.; 7; 51-54, 1982.

14. Vieth, W.R. and Venkatasubramanian, K.; Immobilized cell system, in Enzyme Engineering, PYE, H.E.K. AND WEETAAL, H.H.(eds.).; Plenum, New York; Vol. 4.; p.307-316, 1978.

REMOVAL OF MACRONUTRIENTS FROM WASTEWATERS BY

IMMOBILIZED MICROALGAE

D. Proulx and J. de la Noue

Groupe de Recherche en Recyclage Biologique
et Aquiculture, Universite Laval
Ste-Foy, Quebec, G1K 7P4

INTRODUCTION

Water management is a crucial problem in most, if not all countries. Unfortunately, the primary and secondary wastewater treatment systems currently operating seem inadequate to efficiently treat the huge volumes of nutrient-laden effluents involved. Tertiary treatments are badly needed, but are too expensive. Moreover, the wastage of nutrients caused by the traditional tertiary processes is deplorable, since it is possible to consider these substances as inputs for controlled biological systems for biomass production.

One of the possible biotreatments for converting the nutrients that cause eutrophication (N,P) into useful biomass is solar biotechnology through the cultivation of microalgae (see Shelef and Soeder, 1980). This type of process has been studied for some 40 years now (Caldwell, 1946; Oswald and Gotaas, 1957; Oswald and Golueke, 1968) and offers numerous advantages over tertiary chemical and physico-chemical treatments (Table 1).

Firstly, biological tertiary systems can operate under relatively wide pH variations, whereas chemical systems usually require a defined pH for optimal operation. Secondly, solar biotechnologial treatments do not create secondary pollution problems as do chemical processes employing various chemical flocculants. Thirdly, solar biotechnology can generate algal biomasses that can be used for animal feeding (Kihlberg, 1972; Mokady et al., 1978; Sandbank and Hepher, 1980) or as a source of various chemicals (Aaronson et al., 1980).

302

Parameter	Physico-chemical treatments	Biological treatments
Development	industrial	mostly R & D
Cost	unaffordable	expensive to unaffordable
Sensitivity to factors:		
toxic substances	low	very high
physico-chemical parameters (pH)	high	rather low
Secondary pollutants (Al, Fe, ...)	medium to high	no
Land requirements	low	high
Production of usable biomass or chemicals	no	yes

Table 1. Comparison between physico-chemical and biological tertiary treatment process for wastewaters.

There are, however, some restrictions to the use of microalgal cultures for the tertiary treatment of wastewaters. The first disadvantage is cost which also applies to traditional systems currently too expensive to be widely used. This could be overcome by the selling of usable biomass or, even better, by the extraction of added-value products. The second drawback is that solar systems usually require favourable climatic conditions for maximal efficacy. It has recently been shown, however, that some future developments may allow algal cultures to operate under greenhouse conditions in northern climates (Pouliot and de la NoUe, 1985), which require less tank surface than normal cultures for the same efficacy.

Large-scale solar biotechnology is still confronted with the difficulty of harvesting microalgal biomass, at least in the case small unicellular Chlorophyceae. Various techniques have been proposed or developed, some of them rather costly (ion exchange, electric fields, utrasonic systems) and inefficient. The most promising means of harvesting or retaining the algae in the reactor fall into two broad categories: sedmimentation and filtration (Table 2). The physico-chemical processes are generally efficient but too energy-intensive. Chemical flocculation (Lavoie et al., 1984) is to be avoided because it leads to secondary pollution problems (Table 3), and possible spoilage of the quality of the harvested biomass, with the exception of chitosan flocculation (Lavoie and de la NoUe, 1983, 1084; Lavoie et al.,1984; Morales et al., 1985), which might be an interesting alternative.

With these difficulties in mind, we decided to explore the use of immobilized microalgae. Immobilization eliminates the necessity for harvesting algal cells, and systems operating under semi-continuous or continuous modes can be easily established.

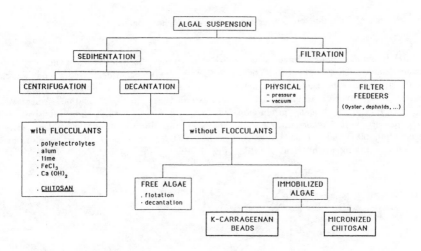

Table 2. Recovery of Microalgae.

Flocculant	Concentration mg L^{-1}	Reference	Remarks
Alum	75-100	McGarry (1970)	Secondary pollution
	125-170	Van Vuuren and Van Vuuren (1965)	
	70	Lavoie et al. (1984)	
Lime	300-400	Van Vuuren and Van Vuuren (1965)	Raised pH. Inefficient. Salt deposits on machinery.
	100	Nigam et al. (1980)	
FeCl$_3$	30	Ives (1959)	Secondary pollution
Cationic polyelectrolytes	50-100	Tilton et al. (1972)	Reported as carcinegenic
Cationic polyamine (Dow C-31)	3	Tenney et al. (1969)	Reported as carcinegenic
Chitosan	50	Nigam et al. (1980)	Non toxic,
	30	Lavoie et al. (1984)	Efficient,
	2	Morales et al. (1985)	Relatively expensive.

Table 3.

Flocculating agents successfully used for microalgal recovery (Adapted from Morales et al., 1985)

Performance of Immobilized <u>Scenedesmus</u> in Carrageenan Beads

Relatively few papers have been published on immobilized microalgae and only a few systems have been tested: production of sulphated polysaccharides (Gudin and Thomas, 1981), production of fuels and chemicals (Baillez et al., 1983; Rao and Hall, 1984), generation of oxygen (Adlercreutz and Mattiasson, 1982), nitrogen fixation (Musgrave et al., 1982) and hydrogen production (Muallen et al., 1983), the last two with <u>Cyanobacteria.</u> It has been shown by Chevalier and de la NoUe (1985a) that cells of <u>Scenedesmus</u> immobilized on k-carrageenan beads (0.5 mm diameter) are as efficient as free cells in taking up nutrients (NH^+_4, Fig. 1; PO^{3-}_4, Fig. 2)

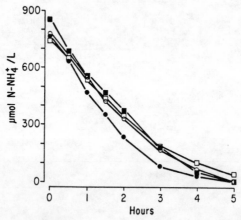

Figure 1. Nitrogen uptake from the secondary effluent by duplicate cultures of free (● , O) and immobilized (■ , □) <u>Scenedesmus</u> <u>obliquus</u> at pH 9.0 (From Chevalier and de la NoUe, 1985a).

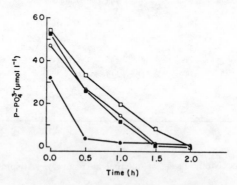

Figure 2. Phosphorus uptake from secondary effluent by duplicate cultures of free (● , O) and immobilized (■ , □) <u>S.</u> <u>obliquus</u> at an adjusted pH of 7.7 (From Chevalier and de la NoUe, 1985a)

from secondary urban effluents. The growth of <u>Scenedesmus</u> was essentially the same whether or not the microalgae were immobilized (Fig. 3). We may conclude that the

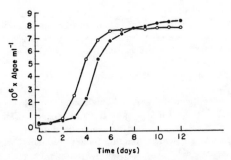

Figure 3. Growth curves of <u>Scenedesmus</u> <u>obliquus</u> in Valcartier secondary effluents. Free (O) and immobilized (●) cells (From Chevalier and de la NoUe, 1985a)

polymer structure of carrageenan did not interfere with growth and that small ions such as NH^+_4 and PO^{3-}_4 could freely diffuse through the gel matrix.

These positive results prompted Chevalier and de la NoUe (1985b) to test the amenability of hyperconcentrated cultures of <u>Scenedesmus</u> <u>quadricauda</u> to immobilization, since such cultures have been shown by Lavoie and de la NoUe (1985) to be efficient for the tertiary treatment of wastewaters. The results obtained with <u>Scenedesmus</u> cells immobilized in carrageen beads were again positive (Fig. 4); the cells were as efficient as free ones for

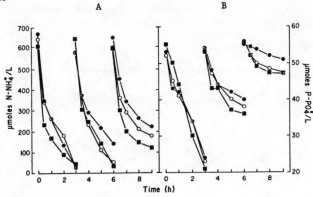

Figure 4. Nitrogen (A) and phosphorus (B) removal from wastewaters with immobilized hyperconcentrated <u>Scenedesmus</u> <u>quadricauda.</u>
● : 0.885 g d.w. L^{-1}; O : 1.89 g d.w. L^{-1};
■ : 3.29 g d.w. L^{-1}. (From Chevalier and de la NoUe, 1985b).

the removal of NH^+_4 and PO^{3-}_4 ions from secondary effluents. It is likely that the semi-continuous system was operated at too high a turnover rate since the efficiency for both the immobilized and free cells decreased with each cycle (Fig. 4). This was especially noticeable with respect to phosphate removal.

These results, although encouraging, do not necessarily allow the proposal of such systems for large-scale wastewater treatment. One of the obvious drawbacks is the cost of carrageenan. It will be necessary to find and to test other, less expensive, immobilization polymers before such a process is to be applied to wastewater treatment.

Chitosan-Immobilized Phormidium

In previous work, we noticed that some small filamentous algae (Phormidium or Lyngbya) attached readily to Daphnia exuviae and to chitosan particles (Proulx, unpublished results). We therefore decided to test the ability of short filaments of Phormidium sp. to exhaust nutrients from secondary effluents when immobilized on chitosan flakes (500 μm particles). The obvious advantage to such a system would be the easy separation of the algae from the culture medium through rapid settling of the particles on stopping aeration.

Industrial grade chitosan (Kypro, Seattle, Wah, USA) was used and full colonization by Phormidium sp. was obtained after 4 weeks. Batch and semi-continuous cultures were then conducted on secondary effluent from the wastewater treatment plant at Valcartier, near Quebec. The initial composition of the culture medium was: $N-NH^+_4$, 9.5 mg L^{-1}; $N-(NO^-_2 + NO^-_3)$, 2.5 mg L^{-1}; PO^{3-}_4, 2.2 mg L^{-1}; total suspended solids, 10 mg L^{-1}; chemical oxygen demand, 40 mg O_2 L^{-1}; pH, 7.0. The experiments were done in 1 L flasks containing 400 ml of medium. The temperature was maintained at 21°C and continuous light was provided at 360 μE s^{-1} m^{-2} Parr.

The algae remained in good condition during the whole experimental period and monospecificity was maintained despite the lack of any precaution, an observation of great relevance to large-scale operation. The final effluent was limpid and free of nutrients, i.e. inorganic nitrogen (NH^+_4, Figs 5 and 6)

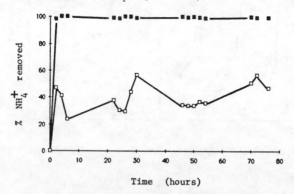

Time (hours)

Figure 5. Removal of NH^+_4 by chitosan-immobilized Phormidium (■4 g d.w. algae L^{-1}) and chitosan alone (□5 g chitosan L^{-1}). ,

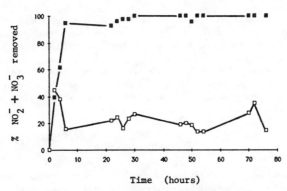

Figure 6. Removal of NO$^-_2$ + NO$^-_3$ by chitosanimmobilized <u>Phormidium</u> (■ 4 g d.w. algae L^{-1}) and chitosan alone (□ 5 g chitosan L^{-1}).

and orthophosphate (Fig. 7). It can be seen that in a short time

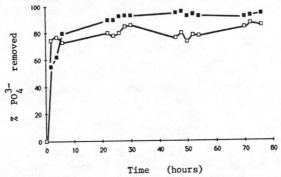

Figure 7. Removal of PO$^{3-}_4$ by chitosan-immobilized <u>Phormidium</u> (■ 4 g d.w. algae L^{-1}) and chitosan alone (□ 5 g chitosan L^{-1}).

(4 - 6 h) over 95% of the nitrogen had disappeared from the culture medium. The pH, which was initially 7.0, rose to between 9.5 and 10.5 over 22 h, typical of active algal suspensions.

To assess the possible participation of chitosan itself in the purification process, through entrapment of ions for example, we used a suspension of concentrated chitosan flakes (5 g L^{-1}) as a control. Under these conditions, 40% of the inorganic nitrogen disappeared over 76 hours. It is likely that most of this nitrogen removal was due to NH$_3$ stripping, especially considering the pH attained (9.5 - 10.5). We cannot, however, rule out the possible retention of some NO$^-_3$ and NO$^-_2$ ions by the chitosan through ionic binding with the positive charges borne by the chitosan molecules. The same explanation may hold for the removal of orthophosphate from the culture medium

(Fig. 7). We can see that 80% of the orthophosphate was removed in the presence of chitosan alone, the presence of the microalgae then being responsible for a further 10% removal over 22 h, the maximum removal measured being 94%. It could well be that most of the PO_4^{3-} was precipitated with Ca^{2+} ions coming from the chitosan, which is known to contain some calcium (Muzzarelli and Pariser, 1978; Colwell et al., 1984), which could be released into solution. Considering the high pH values of the cultures, the likelihood of phosphate precipitation was high. Some of the PO_4^{3-} was, of course, taken up by the cells.

CONCLUSION

The immobilization of microalgae appears to work nicely for either small Chlorophyceae such as Scenedesmus entrapped in carrageenan beads, or for larger filamentous algae such as Phormidium attached to citosan flakes. In both cases, the stability of the system was found to be good, with monospecificity well-assured, and nutrient removal was substantial, even for phosphate in the case of chitosan alone. This latter observation requires further work to elucidate the mechanism involved, especially in view of the fact that it was reproducible with semi-continuous cultures (unpublished results).

ACKNOWLEDGEMENTS

The authors wish to thank the Natural Science and Engineering Research Council of Canada (NSERC) and the Quebec Fund FCAR for financial support.

LITERATURE CITED

Aaronson, S., Berner, T. and Z. Dubinsky. 1980. Microalgae as a source of chemicals and natural products. In: G. Shelef and C. J. Soeder (EAds.) Algae Biomass, Elsevier, North Holland Biomedical Press, Amsterdam, pp 575-601

Adlercreutz, P.and B. Mattiasson. 1982. Oxygen supply to immobilized cells: 1. Oxygen Production by immobilized Chlorella pyrenoidosa. Ensyme Microb. Techol., 4 :332-336.

Baillez, E., Largeau, C., Casadevall, E. and C. Berkaloff. 1983. Effets de l'immobilisation en gel d'alginate sur l'algue Botryococcus braunii. C.R. Acad. Sc. Paris (serie III), 296:199-202.

Caldwell, D.H. 1946. Sewage oxidation pounds- Performance operation and design. Sewage Works J., 18 :433-458.

Chevalier, P. and J. de la NoUe. 1985a. Wastewater nutrient removal with microalgae immobilized in carrageenan. Enz. Microbiol. Technol., 7 :621-624.

Chevalier, P. and J. de la NoUe. 1985b. Efficiency of immobilized hyperconcentrated algae for ammonium and orthophosphate removal from wastewaters. Biotechnol. Leatt., 7 :395-400.

Colwell, R.R., Pariser, E.R. and A.J. Sinskey (Eds). 1984. Biotechnology of Marine Polysaccharides. Proceedings of the Third Annual MIT Sea Grant College Lecture and Seminar. Montreal, McGraw-Hill International Book Co.

Gudin.C. and D. Thomas. 1981. Production de polysaccharides sulfates par un photobioreacteur a cellules immobilisees de Phorphyridium cruentum. C.R. Acad. Sc. Paris (serie III), 293 :35-37.

Ives, K.J. 1959. The significance of surface electric charge on algae in water purification. J. Biochem. Microbiol. Technol. Eng., 1 :37-47.

Kihlberg, R. 1972. The microbe as a source of food. Ann. Rev. Microbiol., 26 428-455.

Lavoie, A. and J. de la NoUe. 1983. Harvesting microalgae with chitosan and economic feasiblity. J. World Maricult. Soc., 14 :685-694.

Lavoie, A. and J. de la NoUe. 1983. Hyperconcentrated cultures of Scendesmus obliquus: a new approach for wastewater biological tertiary treatment? Wat. Res., 15 :1437-1442.

Lavoie, A., de la NoUe, J. and J.-B. Serodes. 1984. Recuperation de microalgues en eaux usees: Etude comparative de divers agents floculants. Can. J. Civil Engineer., 11 :266-272.

McGarry, M. 1970. Algal flocculation with aluminium sulphate and polyectrolytes. J. Water Pollut. Control. Fed., 42 :191-200.

Mokady, S., Yannai, S., Einov, P. and Z Berk. 1978. Nutritional evaluation of the protein of several algal species for broilers. Arch. Hydrobiol./Beih. Ergebn. Limnol., 11 :89-97.

Morales, J. de la NoUe, J. and G. Picard. 1985. Harvesting marine microalgae by chitosan flocculation. Aquacult. Engineer., 4 :257-270.

MUallen, A., Bruce, D. and D.O. Hall. 1983. Photoproduction of H_2 and $NADPH_2$ by polyurethane immobilized Cyanobacteria. Biotechnol. Lett., 5 :365-368.

Musgrave, S.C., Kerby, N.W., Codd, G.A. and W.D.P. Stewart. 1982. Sustained ammonia production by immobilized filaments on the nitrogen-fixing Cyanobacteria Anabaena 27893. Biotechnol. Lett., 4 :467.

Muzzarelli, R.A.A. and E.R. Pariser. 1978. Proceedings of the first international conference on chitin/chitosan, April 1977. MIT Sea Grant Program, Cambridge, Mass.

Nigam, B.P., Ramanathan, P.K. and L.V. Venkataraman. 1980. Application of chitosan as a flocculant for the cultures of the green algae: Scenedesmus acutus. Arch. Hydrobiol., 88 :378-387.

Oswald, W.J. and H.B. Gotaas. 1957. Photosynthesis in sewage treatment. Trans. Am. Soc. Civ. Engin., 122 :73-105.

Oswald, W.J. and C.G. Golueke. 1968. Harvesting and processing of waste-grown microalgae. In: Algae, Man and the Environment. D.F. Jackson (Ed.), Syracuse University Press, Syracuse, NY, pp. 371-390.

Pouliot, Y. and J. de la NoUe. 1985. Mise au point d'une usine-pilote d'epuration des eaux usees par production de microalgues. Rev. Frcse Sci. de l'Eau, 4 :207-222.

Rao, K.K. and D.O. Hall. 1984. Photosynthetic production of fuels and chemicals in immobilized systems. Trends Biotechnol., 2 :124-129.

Sandbank, E. and B. Hepher. 1980. microalgae grower in wastewater as an ingredient in the diet of warm-water fish. In: G. Shelef and C.J. Soeder (Eds.). Algae biomass, Elsevier-North Holland Biomedical Press, Amsterdam, pp. 697-706.

Shelef G. and C.J. Soeder (Eds.). 1980. Algae Biomass, Production and Use. Elsevier-North Holland Biomedical Press, Amsterdam, 852 p.

Tenney, M.W., Echelberger, Jr. W.F., Schuessler, R.G. and J.L. Pavoni. 1969. Algal flocculation with synthetic organic polyelectrolytes. Appl. Microbiol., 18 :965-971.

Tilton, R.C., Murphy, J. and J.K. Dixon. 1972. The flocculation of algae with synthetic polymeric flocculants. Wat. Res., 6 :155-164.

Van Vuuren, L.R.J. and F.A. Van Vuuren. 1965. Removcal of algae from wastewater maturation pond effluent. Water Pollut. Control. Fed. J., 37 :1256-1262.

IMMOBILIZED CELLS IN ANAEROBIC WASTE TREATMENT

Ralph A. Messing
168 Scenic Drive South
Horseheads, New York 14845

ABSTRACT

Both fuel and pollution concerns have occupied the thoughts of the current generation. A variety of approaches have addressed one or the other of these concerns; however, if an approach is offered that could, potentially, solve both problems simultaneously, it should be evaluated expeditiously. This truly is the case of anaerobic waste treatment with immobilized microbes.

Massive populations of microbes can be immobilized in the pore structure of controlled-pore ceramics. The cells are closely packed in the pores because they immobilize on their minor dimension. The rapid and continuous delivery of nutrients and the removal of metabolic wastes promote early cell division with the production of uniform mini-cells. This is not unlike events that occur in nature.

A two-stage reactor was designed to exploit the advantages of immobilized microbes for methane production and waste treatment. The first stage contained immobilized acid-formers while the second stage contained, primarily, immobilized Methanobacter species.

A three-fold approach for retaining CO_2 in the liquid phase was employed to increase the efficiency of its reduction to CH_4. This approach included the elevation of pH and pressure and the reduction of temperature in the second stage.

KEYWORDS

Waste treatment, anaerobic, bioreactor, immobilized microbes, methane, continuous reactor, pollution, fuel.

311

312

INTRODUCTION

We must not allow ourselves to become complacent as a result of the glut of oil on the world market. This sitution is but temporary. the twin problems of pollution and the development of reasonable and manageable alternative energy sources burden our future. These two problem are inexorably wedded. The recent catastrophe at the Chernobyl nuclear reactor drives this point home. Nuclear energy is difficult to manage at this time because of the potential for a plant upset and the disposal of spent nuclear fuel. Both of which terminate in the pollution of the air and the water. Coal offers no solace. Its mining and processing pollute the water while its burning pollutes the air. The treatment of waste to avoid pollution problems is of greater import with increasing population. Communities, formerly, disposed of their wastes in rivers, lakes and the oceans. Prohibitions for this behaviour have now been established. The alternative disposals of landfills and surface spreading are now in disrepute. The problems arising from the latter techniques are diminished land availability and ultimate ground water pollution. In addition to agricultural, food, processing and human wastes, there is a great concern with respect to toxic wastes. Pyrolysis appears to be a reasonable approach to disposing of these wastes; however, this process becomes infeasible in dilute solutions because of the energy and equipment demand required for water removal. A more appropriate alternative to massive liquid waste treatment is that of high rate aerobic treatment; however, this process requires great expenditures of energy or agitation and oxygen delivery. In addition, no fuel is evolved, and large quantities of biomass that must be disposed are produced. Anaerobic waste treatment, on the other hand, did not represent a feasible approach to treating large volumes of waste solutions, although it is modestly fuel efficient and produces considerably less biomass, because it was a time consuming batch process which was subject to upsets.

The advent of the anaerobic filter(1) markedly increased the potential for applying anaerobic waste processing to treating large volumes of waste solutions. The anaerobic filter process employs an upward flow delivery on a continuous basis of a waste stream through a bed of rocks containing an immobilized film of microbes on their surface in an enclosed vessel. The gas, a potential fuel, is separated from the treated effluent liquid stream. The anaerobic filter markedly reduced the anaerobic waste processing time from greater than 30 days required by the traditional batch equipment to a day or less with this continuous system. The anaerobic filter does not appear to be as sensitive to upset as the batch reactors. Although the anaerobic filter produces a fuel, biogas, yields less biomass and requires less energy than does the anaerobic processes, associated problems remain to be solved.

Biogas is a mixture of methane, carbon dioxide and hydrogen sulfide. The fuel value derived from methane is dimished by the presence of carbon dioxide while hydrogen sulfide is responsible for the corrosion of equipment. Both carbon dioxide and hydrogen sulfide are removed by scrubbing in order to produce a desirable fuel from biogas. The biogas produced with the batch reactor generally contains 50% CH_4, 49% CO_2 and about 1% H_2S. The anerobic filter produces a superior gas composed of about 70% CH_4, 29% CO_2 and somewhat under 1% H_2S: unfortunately, it too must be scrubbed.

Another problem encountered in the anaerobic filter is the trade-off between rock size and the ability to handle particles that are suspended in the waste stream. Large rocks will allow the processing of suspended particles of significant size and quantity; however, the surface area of the rocks for the immobilization of the microbial film per unit volume of reactor will be very low, and thus the reactor will be inefficient. The surface available for the immobilization of the active microbes may be increased by employing small rocks, but then high pressure drops and restriction to flow are encountered as result of the packing of particles in the void volume.

IMMOBILIZED CELLS IN CONTROLLED-PORE CERAMICS

The obvious path to increase the number of cells immobilized on surfaces within a given volume is to use the third dimension or the internal surfaces of a porous support. The internal surface area of a porous support is several orders of magnitude greater than the external surface. When the relationship between the size of the microbe and the pore diameter of the support(2) was examined, it was noted that the greatest available surface area for microbes that reproduce by fission in the rapid growth phase was present when the pore diameters were five times larger than the major dimension of the cell. A model was constructed, Figure 1, based on the empirical data. The rationale for this model was that if cells immobilize on their minor dimension on opposite walls of the pore and they reproduce by fission, then at least four cell lengths would be required for reproduction; however, another cell length would be required for continuous reproduction order to remove the progeny.

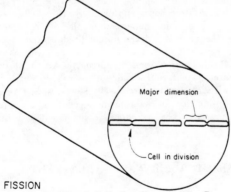

FISSION

Figure 1. Model of Immobilized Cells for Growth in a Pore

If the pore diameter is five times the major dimension of the cell, then approximately 10^9 cells can be immobilized per gram of support material, or between one and two magnitudes more cells can be immobilized than on the non-porous equivalent of that support material. Additional support for the model appeared several years later in the form of scanning electron micrographs of cells immobilized on controlled-pore alumina, Figure 2.

Figure 2. SEM of cells immobilized on controlled-pore alumina.

The scanning electron micrograph clearly indicates that all cells are immobilized on their minor dimension. The cells are packed closely together on the surface. Far more cells can be immobilized on their minor dimension than either on their major dimension or randomly. Cells will immobilize on their minor dimension if the nutrients are offered by the bulk solution and not the support.

The value of this immobilization process was demonstrated(3) by placing one gram of immobilized E. coli K 12 composite (10^9 cells) in a small column and operating it as a plug-flow reactor by delivering nutrients rapidly while removing metabolic wastes. At flow rates of between 1 and 22.5 liters/hour, approximtely 10^{11} cells were delivered in the effluent stream per hour. The major dimension of the effluent cells was between 0.65 and 0.85 microns while the original culture immobilized contained cells between 1 and 6 microns. If these mini-cells are allowed to remain in the medium for more than 1 hour, they exhibit cell lengths between 1 and 6 microns. At the highest flow rates possible, the original immobilized parent cells could not be washed out. Apparently, the rapid delivery of nutrients with concomitant removal of wastes and progeny promotes early cell division and the production of uniform mini-cells. In addition, if the 10^9 parent cells produced 10^{11} progeny cells per hour, each parent cell delivered a new progeny every 36 seconds(4).

IMMOBILIZED CELL BIOREACTOR FOR WASTE TREATMENT

The value of immobilized cells was demonstrated in a two-stage reactor, figure 3, (3,5) for the anaerobic treatment of wastes. This bioreactor is an improvement upon the original anaerobic filter in that it separates two markedly different populations, the acid-formers and the Methanobacter, in order to improve their individual requirements and to improve the utilization of carbon. The rationale for the design of the reactor was based upon observations of natural occurrences. Natural gas from shallow wells is the product of anaerobic fermentation. That gas, generally, contains 90% methane or greater

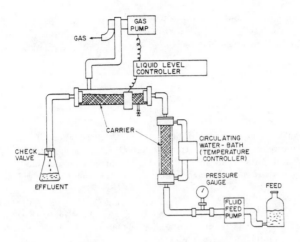

Figure 3. <u>Two-stage anaerobic waste treatment bioreactor</u>

and less than 5% CO_2 while the gas evolved fromm waste treatment processes contains between 40 and 75% methane and 25 to 60% CO_2. Under natural conditions, the acid formers which require the presence of some oxygen are located at or close to a surface interfacing with air. The acid-formers break down large molecules (proteins, cellulose, fats, etc.) by oxidative and hydrolytic processes to small molecules (four carbon chain and less) while consuming oxygen. On the contrary, <u>Methanobacter</u> are located at considerable depth, remote from a source of oxygen. The <u>Methanobacter</u> are obligate anaerobes which are inhibiting or are inactive in the presence of even low concentrations of oxygen. This group of microbes is able to metabolize only molecules that are four carbon chains or less. It would appear from these observations that <u>Methanobacter</u> require pressure and cooler temperature than do the acid-formers. Our studies (3,5) of the anaerobic degradation of sewage and manure with immobilized acid-formers in the first stage and immobilized <u>Methanobacter</u> in the second stage of the bioreactor, did, in fact, indicate that when we applied pressure to the <u>Methanobacter</u> by inserting a check-valve in the effluent passage of the second stage and when we operated the second stage at a lower temperature than the first stage we could produce an improved performance. Approximately 45% of the total influent carbon was converted to methane, and the evolved gase contained greater than 90% methane and less 5% CO_2 at residence times of four hours in the case of sewage. This improved performance may be attributed to an increased retention of CO_2 in solution with increased pressure and decreased temperature which, in turn, leads to more reduction of that gas to methane by the microbes that are located in the aqueous phase.

It has been estimated that between 45 and 55% of the total carbon found in sewage is required for cell growth; therefore, the immobilized cell bioreactor was responsible for converting 100% of the carbon available for the production of methane to that gas.

In addition to sewage and manure, wine stillage, pulp and paper waste liquors and pumpkin wastes have been processed with varying degrees of success. This bioreactor appears to be limited to processing waste streams that contain less than 40,000 PPM COD.

CONCLUSION

In order to make a positive contribution to energy production, it has been estimated that this bioreactor requires a waste stream of greater than 5,000 PPM COD. Although sewage usually contains less than 1,000 PPM COD, this should not inhibit its application to this waste since there is a marked reduction in the energy required to operate this process.

Our current efforts are directed to a modified two-stage bioreactor that we believe will be capable of processing streams containing 100,000 PPM COD. In addition, we are evaluating better supports for the immobilization of cells.

REFERENCES

1. J.C. Young and P.L. McCarty. The anaerobic filter for waste treatment. J. Water Pollut. Control Fed. 41, R160-R173(1969).

2. R.A. Messing and R.A. Oppermann. Pore dimensions for accumulating biomass I. Microbes that reproduce by fission or by budding. Biotechnol. Bioeng. 21, 49-58(1979).

3. R.A. Messing and T.L. Stineman. Rapid production of methane with immobilized microbes. Annals of The New York Academy of Sciences 413, 501-513(1983).

4. R.A. Messing. Bioreactors and immobilized cells. SIM News 35 (#3), 4-7(May 1985).

5. R.A. Messing. Immobilized microbes and a high-rate, continuous waste processor for the production of high Btu gas and the reduction of pollutants. Biotechnol. Bioeng. 24, 1115-1123(1982).

METHANE PRODUCTION IN IMMOBILIZED CELL BIOREACTOR

J. M. Scharer, A. Bhadra and M. Moo-Young
Department of Chemical Engineering
University of Waterloo
Waterloo, Ontario, N2L 3G1

SUMMARY

Methanogenesis from organic acids was investigated in immobilized cell (downflow stationary fixed film) bioreactors. The biomass support materials included ceramic Raschig rings, hardwood (beach) chips and charcoal. The performance of the biofilm on the various supports was evaluated using acid products from paper mill sludge and synthetic acid mixtures. Wood ships were the most effective support material; a maximum methane productivity of 3.95 m^3 (CH$_4$)/m^3/day was obtained at a nominal retention time of 0.78 days. Acids conversion efficiency was 68.6%. The conversion efficiency improved with higher retention times (85.3% at 1.3 days), but the productivity declined (2.76 m^3/m^3/d).

The adsorption of acetogenic and methanogenic bacteria to the support material (wood chips) has been assessed in terms of adsorbed particulate organic nitrogen (PON). Generally, the surface concentration of cell mass was directly related to methane generation. The adsorbed PON/m^2 of support could be related to the cell concentration in solution by a Freundlich-type isotherm. The biomass growth yield (Y_g) and the maintenance coefficient (m) were estimated to be 0.124 g cells/g acid consumed and 0.107 g cells/g acid consumed/day, respectively.

INTRODUCTION

The advantages of fixed film bioreactor configurations for methane generation are well established (1,2). The application of cell immobilization becomes particularly attractive in case of two-stage anaerobic digestion of solid wastes where the first step, acidogenesis, is carried out in a separate reactor and the effluents of the first reactor are used as substrate for methane production. Using this method, the immobilized cell reactor can be operated at much higher throughput (loading and flow rate) and no problems with plugging and mixing requirements arise.

Our work involved such a two-stage anaerobic digestion system for solid wastes, e.g. pulp mill sludge, corn stover and animal excreta. In the first stage, the digestible solids were converted to volatile fatty acids in a semicontinuous reactor operating at either mesophilic (37°C.) or thermophilic (55°C.) temperature. The acid mixture product contained acetic, propionic, butyric and occasionally isovaleric acid. A downflow stationary fixed film (DSFF) reactor was used as a second stage to perform the acetogenic and methanogenic steps to produce methane and carbon dioxide. This paper reports the use of three different support materials: hard woodchips (beach), sized charcoal and ceramic Raschig rings. Both synthetic acid mixtures and volatile fatty acids obtained from pulp mill sludge were used as feed stock.

MATERIALS AND METHODS

Fixed film reactors of either 2.5 L or 14 L capacity were used in the present study. The schematic diagram of the two stage system is shown in Figure 1. The operation and details of the reactors have been reported in our earlier communication (3).

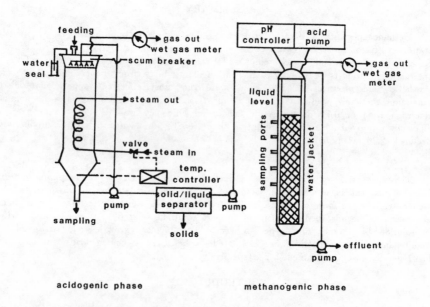

Figure 1. schematics of the two-stage anaerobic digestion apparatus

The volatile fatty acids concentration in the broth was determined by gas chromatography (Hewlett-Packard Model 5880A) as reported previously (4). Gas compositions were also determined chromatographically (Hewlett-Packard Model 700) (3).

RESULTS AND DISCUSSION

The effect of pH on methane production was studied by controlling the pH at different levels (from 5 to 9). As shown in figure 2, the optimum was pH 6.8, but no

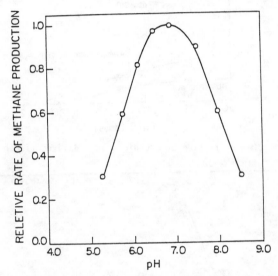

Fig. 2:   Effect of pH on methane generation.

significant differences in the rate of methane production occurred between pH 6.5 and pH 7.5. The utilization of acetate, propionate, and butyrate mixture is shown in Figure 3 at pH 7.0 (controlled). Acetic acid was consumed very rapidly in comparison to the other two acids. Initially, propionic and butyric acid were converted slowly but as the concentration of acetic acid declined, the consumption of these acids increased significantly. The propionic and butyric acids are known to be degraded to acetic acid by acetogenic bacterial before methanogenesis takes place. The presence of acetic acid has an inhibitory effect on the utilization of either propionic or butyric acid. As shown in Figure 4 and Figure 5, at significant concentrations of acetate, the rate of utilization of either propionate or butyrate, was inhibited. A competitive type inhibition model has been proposed and tested by Girard et al. (5).

320

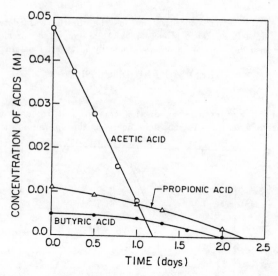

Fig. 3. Utilization of organic acids in fixed film bioreactor.

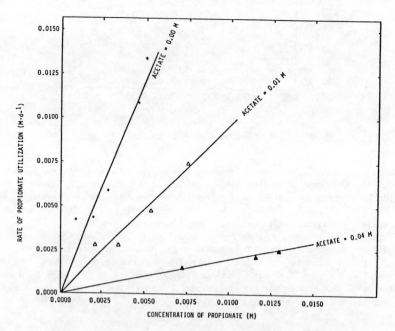

Fig. 4. Effect of acetate on utilization of propionic acid.

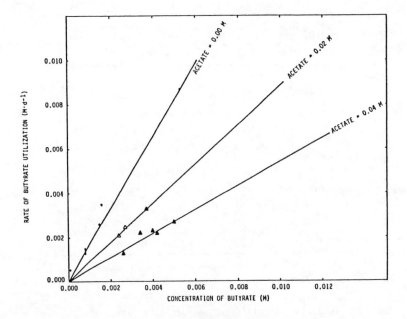

Fig. 5. Effect of acetate on utilization of butyric acid.

The three support materials: woodchips, charcoal and Raschig rings, were compared for their performance in terms of acid conversion, gas yield and productivities (Table 1). Both woodchips and charcoal were found to be reasonably good support materials, whereas Raschig rings were definitely inferior. The conversions of individual acids are presented in Table 2. The utilization of acetic acid, which was the major component of the feedstock, was always essentially complete with either woodchips or charcoal as support material. As shown in Table 1, the gas yield and productivity values were higher in case of woodchips due to the more efficient conversion of the higher acids. This may imply, that the attachment or the activity of acetogenic bacteria was enhanced by the use of woodchips.

Support Material	Surface area volume m^2/m^3	Input acid conc gL^{-1}	RT d	Conversion of acids %	Gas yield l gas/ g acid consumed	Methane Productivity l/l/d
		5.14(a)	1.04	97.0	0.64	2.15
		6.81(a)	1.30	93.75	0.64	2.33
Woodchips	302	9.13(b)	1.30	85.34	0.63	2.76
		9.13(b)	0.78	71.40	0.62	3.56
		10.54(b)	0.78	68.59	0.62	3.95
		5.14(a)	1.25	91.16	0.62	1.63
Charcoal	315	7.33(a)	1.25	82.16	0.65	2.57
		9.13(b)	0.85	69.29	0.62	3.16
Raschig rings	243	11.67(b)	1.04	39.6	0.67	2.08

(a) Substrate obtained by acidogenesis of pulp mill sludge.

(b) Synthetic acid mixture.

Table 1. Comparison of different support materials.

Support Material	RT d	Input Acid Conc. gL^{-1}					Conversion of Acids, %				
		AA	PA	BA	IVA	TVA	AA	PA	BA	IVA	TVA
	1.04	2.1	0.99	0.96	T	5.14	100	95.2	94.6	-	97.0
Wood-chips	1.30	3.15	1.2	1.15	0.1	6.81	100	90.5	89.5	100	93.75
	1.30	4.51	1.51	1.45	-	9.13	98.1	75.4	70.7	-	85.34
	1.25	2.1	0.99	0.96	T	5.14	100	89.3	81.4	-	91.16
Charcoal	1.25	2.15	1.15	1.14	0.85	7.33	100	82.2	78.1	63.5	82.16
	1.25	4.51	1.51	1.45	-	9.13	96.5	64.7	58.5	-	78.64

T = trace

Table 2. Conversion of volatile fatty acids using woodchips and charcoal as support material.

323

The effect of retention time and initial acid concentration on the rate of acid utilization as a function of the retention time and influent acid concentration is shown in Figure 6. The rates were calculated as time-averaged values over a period of 10 days for each retention time. For a particular retention time, the rate of acid utilization was higher when the influent acid concentration was higher. It should be noted that, although the rate was higher at lower retention time, the total amount of acid utilized was more at higher retention regardless of the initial acid level.

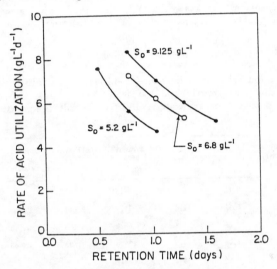

Fig. 6. Effect of S_0 on rate of acid utilization.

In case of slowly growing organisms, the growth and maintenance requirements are very important factors (6). These parameters provide means of assessing the efficiency of the microoganisms for desired product formation. In any aerobic system, the extent of growth at the expense of the carbon source is much less than during aerobic growth, since the ATP yield under anaerobic conditions is substantially less.

The true growth yield (Y_G) and maintenance coefficient have been computed by utilizing substrate material balances (7). These can be represented by the following equation:

$$\left(\frac{ds}{dt}\right)_{Total} = \left(\frac{ds}{dt}\right)_{Growth} + \left(\frac{ds}{dt}\right)_{Maintenance} \tag{1}$$

After proper substitution, equation (1) is transformed to the following form:

$$\frac{1}{Y_{x/s}} = \frac{1}{Y_G} + \frac{m}{\mu} \tag{2}$$

where

$Y_{x/s}$ =   growth yield

μ =   specific growth rate

The values of apparent growth yield, $Y_{x/s}$, and the specific growth rate, μ, are presented in Table 3. The growth parameters (Y_G and m) were computed by plotting

$\dfrac{1}{Y_{x/s}}$ vs. $\dfrac{1}{\mu}$ (Fig. 7).

The slope gave the value of maintenance coefficient (m = 0.107 d^{-1}) and from the intercept the value of true growth yield was found to be 0.124 g.cell/g acid. A low value of Y_G ensures that more substrate is diverted towards non-growth functions, e.g. product formation.

mg Cell in 50 ml solution	Total acid conc, gL^{-1}	mg N g support	% removal of acid	Cell yield $Y_{x/s}$	Specific growth rate, μ
4.75	9.11	0.7	11.6	0.12	0.39
8.0	6.52	2.1	36.7	0.114	0.20
10.25	3.89	3.6	62.2	0.113	0.113
11.25	1.05	4.9	89.8	0.103	0.066
11.90	0.54	5.2	94.7	-	-

Note: The N content of dry cell mass was considered to be 12.4% (W/W).

Table 3. Amount of cell mass in adsorbed and free state.

Adsorption isotherms are useful plots for understanding the adsorption of molecules on solid surfaces. Analogous isotherms can be obtained in case of immobilized cells on solid support material by plotting the amount of cells adsorbed against the concentration of free cell mass in solution. The adsorption isotherm of mixed population of acetogenic and methanogenic bacteria on woodchips was studied in small flasks containing 50 ml of acid solution and 1 g of support material. The amount of cell mass adsorbed was determined by measuring the increase in particulate organic nitrogen (PON) content of the solid. It was found that the removal of acid was directly proportional to cell growth (Table 3). The adsorption isotherm could be represented by the empirical equation proposed by Freundlich (8):

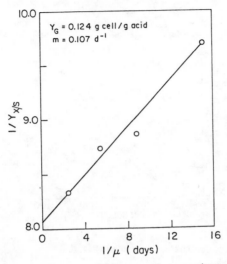

Fig. 7. Evaluation of the growth parameters.

$X = KC^n$

where

$X = $ mg cell/m² support material

$C = $ mg cell/L in solution   (3)

K and n are empirical constants. The numerical values were found to be 2.03 x 10^{-3} and 2.422 respectively for this particular case at 35° C. As shown in Figure 8, the amount of cells adsorbed increased with the increase in cell mass concentration in the solution until a maximum value is reached. This occurs at the maximum possible cell concentration in solution. It is also clear from the nature of the curve that the rate of adsorption is always positive till some maximum value is reached. The final plateau is believed to be due to the formation of a cell layer of finite thickness at the maximum cell concentration in solution.

A comparative study utilizing one-stage and two-stage digestion was performed with pulpmill sludges. The digestions were carried out at mesophilic temperature (37° C.). In the one stage process, the maximum conversion of cellulose (76%) was achieved in 25 days. In two-stage digestion, the retention time in the first stage varied between 6 and 8 days for maximum conversion of cellulose, depending on loading. The overall performance of the two processes is shown in Table 4. The gas yield increased about

326

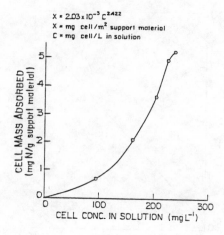

Fig. 8. Sorption isotherm for woodchips.

PARAMETER	ONE STAGE	TWO STAGE
TEMP	37°C	37°C
RT,d	25	9 (7.5 + 1.5)
Conversion of cellulose, %	76	91
Gas Yield l gas/kg solid digestion	0.31	0.55
Productivity l gas/l/day	0.19	0.63

Table 4. COMPARISON OF ONE AND TWO STAGE DIGESTION
OF PULPMILL SLUDGE

1.8-fold and the productivity 3.3-fold in case of the two-stage digestion, which clearly demonstrates the greater efficacy of the two-stage process. The high productivity was due to lower retention time in the second stage which resulted primarily from the higher cell density in the fixed film bioreactor.

CONCLUSION

High rate fixed film reactors were found to be very efficient to produce methane from organic acids when woodchips or charcoal were used as support materials. Conversion of acetate was usually complete, while higher acids were only partially utilized. The presence of acetic acid inhibited the conversion of higher acids. Cell sorption to woodchips was characterized by a Freundlich-type isotherm.

327

Two-stage anaerobic digestion could be successfully applied to convert pulp mill sludges to methane. Two-stage bioconversion was more efficient in comparison to single stage digestion in terms of substate conversion and gas productivity.

REFERENCES

1. S. J. B. Duff and L. van den Berg. Biotechnol. Letters 4, 12, 821 (1982).

2. K. J. Kennedy and L. van den Berg. Agric. Wastes 4, 151 (1982).

3. A. Bhadra, J. M. Scharer and M. Moo-Young. Methanogenesis from volatile fatty acids in downflow stationary fixed film reactor. Biotech. & Bioeng. Accepted (1986).

4. A. Bhadra, J. M. Scharer and M. Moo-Young. Thermophilic bioconversion of cellulose to volatile fatty acids. The Chemical Eng. J. 33, B19-B25 (1986).

5. P. Girard, J. M. Scharer and M. Moo-Young. Two-stage anaerobic digestion for the treatment of cellulosic wastes. The Chemical Eng. J. 33, B1-B10 (1986).

6. S. J. Pirt. Principles of Microbe and Cell Cultivation. Blackwell Scientific Publication, Oxford, 1975.

7. J. E. Barley and D. F. Ollis. Biochemical Engineering Fundamentals, McGraw-Hill, 1986.

8. O. A. Hougen, K. M. Watson and R. A. Ragatz. Chemical Process Principles. Part I. Material and Energy Balances, John Wiley and Sons, Inc., 1958.